A GARDENER'S GUIDE TO

Orchids and Bromeliads

Editors: Sara Rittershausen and Mike Pilcher

Series Editor: Graham Strong

MURDOCH BOOKS

CONTENTS

LEFT: The extraordinary pouched blooms of the paphiopedilums or 'Slipper Orchids' make a bold display with their stately, upright stems.

GROWING ORCHIDS

Orchids make up one of the largest families of all the world's flowering plants. They occur on every continent except Antarctica and have a most remarkable diversity of habitat, form and colour. While the greatest number of orchids is found in tropical and subtropical regions, they can also be found in near-desert conditions, tundra and in mountain country.

Orchid flowers vary in size from large, showy blooms the size of saucers to tiny treasures a couple of millimetres across. Some types of orchid are truly breathtaking in their beauty while others may be quite strange and almost ugly to some eyes, but all are simply fascinating. All orchids should have the spent flowers cut off once they are past their best. Due to the vast diversity of the orchid family, there can be orchids flowering in any season of the year. Some will adhere strictly to their season whereas others may bloom intermittently throughout the year.

LEFT: The fantastically flamboyant flowers of this Laeliocattleya *Quo Vadis 'Floralia' are sure to brighten up any display.*

TYPES OF ORCHIDS

The vast majority of orchids in cultivation are epiphytes that grow on trees, but they use the tree as support only – they are not parasites. Other orchids are lithophytes that grow on rocks, or terrestrial types that grow in the ground. Orchids are further distinguished by the way they grow: monopodial orchids, mainly epiphytes, grow with a single stem and produce aerial roots, while sympodial orchids have a rhizome (running root) that produces a pseudobulb from which growth emerges. Many sympodial orchids are terrestrial. Orchids from cooler and more temperate regions are terrestrials but they can occur in warm regions too. Epiphytic orchids are found only in warmer areas.

HYBRIDS

Orchids in the wild hybridize occasionally, and so natural hybrids arise. In their natural situations orchids are pollinated by a range of creatures, including bees, wasps, birds, bats and beetles. Today, commercial growers and enthusiasts making deliberate cross-pollinations are responsible for introducing many new orchid cultivars and varieties each year. There are now at least 100,000 registered orchid hybrids. Since the 1890s all orchid hybrids have been registered in what is now called the Orchid Hybrid Register run by the Royal Horticultural Society in London. This was previously known as Sanders' Orchid Hybrid Lists, named after the orchid enthusiast who began the daunting task of documenting the whole range of orchids and their parentage.

ANATOMY OF AN ORCHID

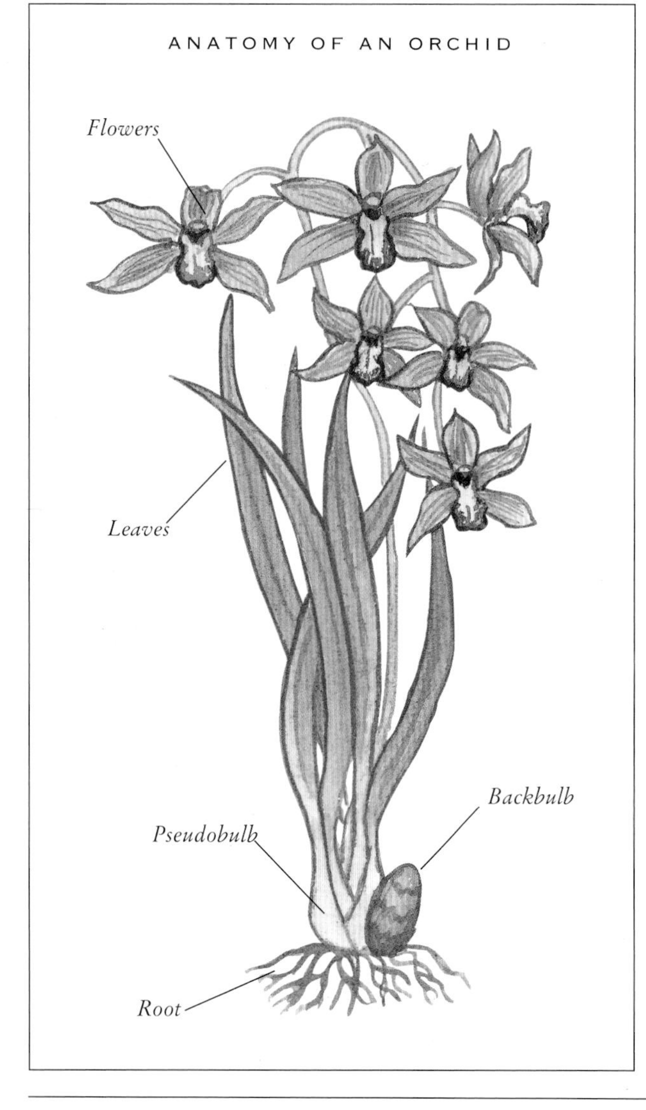

ORCHID CULTURE

For many years orchid culture was strictly the province of the rich and powerful, not only because of the cost of actually acquiring the plants but also because of the cost of building and heating structures for their successful cultivation. Nowadays special prize-winning plants are still expensive but the advent of plant tissue culture, which allows large numbers of plants to be propagated from very little material, has meant that plants are generally more affordable. There is now a huge number of enthusiasts growing orchids all over the world.

Today many orchids are cultivated commercially for the cut-flower trade. Apart from that, only the vanilla orchid, *Vanilla planifolia*, has commercial value. Its flowers grow on a vine and it is cultivated in tropical regions for its aromatic bean, used widely to flavour food.

BUYING ORCHIDS

For a long time the main source of orchids was plants collected in the wild. Now, with loss of habitat and deforestation of many of the world's tropical forests, it is important to conserve and protect what remains. Plants have been collected to the point of extinction in many regions and trade in endangered species is monitored by an international body, CITES or the Convention on International Trade in Endangered Species.

Orchids for sale in specialist or other nurseries are largely cultivars that have been multiplied and grown by commercial growers. If you are starting an orchid collection, you would be better off keeping to the hybrids. These are more plentiful, usually with bigger blooms, and are replaceable if accidentally killed. The species should only be grown by the skilled specialist.

Specialist nurseries have staff who are generally very knowledgeable and eager to help beginner growers, as well as catalogues to help with your choice of plants. Buy your plants while in bloom or choose ones you have seen in other people's collections or in catalogues. Choose the orchid that will give you the greatest pleasure.

Some orchid groups come in a wide range of colours and have an extended flowering season, and you may want to stay with certain colour tones or concentrate on plants that flower at different times of the year. Do not let other people influence your choice.

Once you are 'hooked' on orchids consider joining an orchid society as you will learn a large amount about the plants from other members. You will also have the chance to buy or exchange plants at the society's meetings.

GROWING CONDITIONS

Shade and ventilation

Orchids are not necessarily more difficult to grow than other plants if you give them the right climatic and cultural conditions. Many cool-growing orchids come from high mountain regions with frequent cloud and mist and they will not thrive in the tropics. Likewise, tropical orchids will die quite quickly if not given adequate heat and humidity. Some varieties can be grown in the open in summer. Often the dappled shade from taller trees is enough to provide good growing conditions, or you can build a shadehouse for protection from sun and wind.

Keeping orchids indoors

Orchids are becoming increasingly popular as houseplants, partly due to their high profile but also as modern hybrids, with greater tolerance and colour ranges, are developed.

Modern homes often have a dry atmosphere, but as orchids prefer a humid environment, some moisture should be provided for them. By grouping orchids together or by growing them with other plants that like the same conditions, such as bromeliads and ferns, you can create a suitable microclimate for the plants. You can also stand your plants on humidity trays containing gravel or porous clay pellets. These retain moisture when plants are watered and gradually release it through evaporation, increasing the humidity. However, you should not just rely on the humidity trays to water your orchids – if roots take up too much water through the bottom of the pot they can rot. Instead, take the plants to a sink and water there, letting it flow through the pot and then leave to stand until fully drained before returning to the trays. Another way of creating the right atmosphere is to mist the foliage of the plants regularly with water, especially in warm conditions; this also helps to keep leaves dust-free.

Although orchids do not enjoy direct summer sunshine, good light must be provided if growing indoors. Place the plants near a window where they will get good light but not bright sun, which could scorch their leaves. If the plants are not getting enough light the leaves may become elongated and dark green in colour, and they may even grow towards the light, becoming top heavy. Provide shade on a south-facing windowsill with a net curtain or a piece of greenhouse shade cloth to protect the plants.

Orchids enjoy a variety of room temperatures – some cool, some warm, some in between – so the correct room should be chosen for your plants. If the room is heated most of the time, the temperature not dropping below 15°C (60°F), then this is a warm climate. Although warm loving orchids will like this, do not place the plants too close to the heat source or else there is a danger of them overheating. Cooler growing orchids would prefer an unheated room indoors where the temperature drops to 10°C (50°F) – if kept too warm flowering can be restricted.

Conservatories

A conservatory can make a good environment in which to keep orchids. Grow them together with other plants to create the right humid, shaded conditions.

Spraying water and misting is easier in a conservatory than indoors, providing a higher level of humidity and a better environment to grow the more challenging types. If possible, spray the floor with water daily, especially in warm weather. This will evaporate throughout the day, saturating the air. You could also make a water feature, such as a pool or waterfall, around which you can grow orchids and other moisture-loving plants.

Some heat may be required in the conservatory during the winter, as well as extra shading and ventilation in the summer. The same rules apply as with growing indoors regarding the temperature ranges; cool 10–20°C (50–68°F), intermediate 12–25°C (52–77°F), warm 15–25°C (60–77°F). Try not to mix orchids that need different temperatures.

The blooms of this Vanda x Ascocentrum *hybrid are perfectly shaped and finely speckled in red with rose pink margins.*

Shadehouses

If you have more than a few pots of cool orchids you may decide to house them in a shadehouse for the summer. You can make a simple structure of treated timber or galvanized piping covered with wooden or metal laths. Synthetic shadecloth providing 50 per cent shade is good, although a higher degree of shading may be required for some orchids. Laths or shadecloth will cut down the force of strong winds without making the atmosphere stagnant.

Greenhouses

Greenhouses should have roof vents that can be opened to allow hot air to escape and side vents to draw in cooler air. In very warm conditions wall or ceiling fans can keep the air moving if the humidity and heat become too high. In summer you will also need to provide shading.

Bringing orchids indoors

Many orchids can be brought into the house when in flower. For those with a short flowering period this is fine, but if the flowers last six to eight weeks it is best to cut off the flowers and enjoy them in a vase after about four weeks indoors, to avoid a setback in the plant's growth.

GROWING METHOD

Containers

Orchids can be grown in most types of containers, including terracotta and plastic pots, and baskets of wooden slats or wire. Most like to be grown in pots just

PROPAGATING A CYMBIDIUM ORCHID

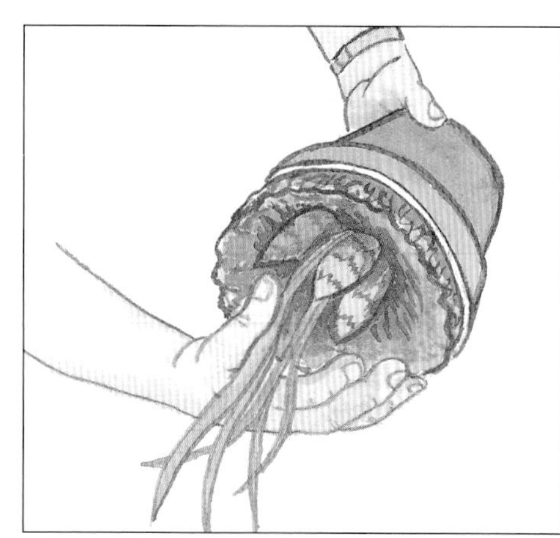

1. Gently ease the plant out of its pot. Run a knife blade around the inside rim if necessary.

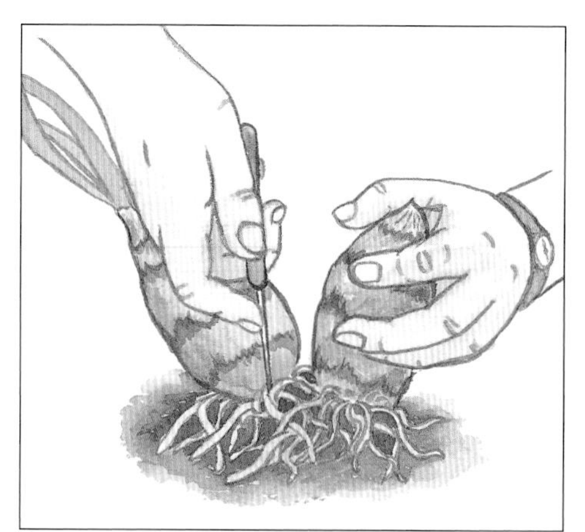

2. Use a sharp knife or secateurs to sever the old leafless backbulbs from the younger growths.

3. Remove the old roots and leaf bases from the backbulb and pot it up. Label it clearly.

4. After several weeks or months a leaf shoot appears. It should flower in three years.

large enough to contain their roots. Many epiphytes will have aerial roots that will grow outside the container.

• Plastic pots are the most commonly used because of their relatively low cost, light weight and ease of cleaning.

• Terracotta pots look attractive and have the advantage of 'breathing', that is, allowing better aeration of the mix inside the pot. They are, however, more expensive, are heavy even before planting, dry out more quickly and their porous surface can allow the growth of algae and slimes and the build-up of fertilizer salts.

• Hanging baskets are essential for some orchids, such as stanhopeas, as their flower spikes grow downwards. The baskets must be lined with soft material, such as coconut fibre, so that the flower spikes can penetrate it.

It is a good idea to have all your orchids in the same type of container, either plastic or terracotta. This will help you to settle on a watering programme and looks more professional.

Mounting an orchid on bark

Some orchids are happiest attached to slabs of cork or tree fern fibre. To grow on bark, either choose an orchid already established on a piece of cork, or transfer one from a pot, picking a plant that has a suitable habit. If the plant has a long, creeping rhizome between the pseudobulbs and is already climbing out of its pot then it is ideal. Plants with tightly packed pseudobulbs do not grow easily on bark.

Potting mixes

There are many types of mix for the cultivation of orchids, but they all have one thing in common: they are fast draining and provide good aeration for plant roots. Bark is the basis of most orchid composts, usually pine bark in various sizes and grades. It is used alone for some orchids or it may be mixed with charcoal, gravel, perlite, peanut shells, spent mushroom compost, coarse river sand, coir peat and sometimes blood-and-bone or slow release fertilizer. Sometimes a mix of only two items, such as pine bark and charcoal, is used. In most cases the potting mix or growing medium is simply there to anchor the roots of the plant. Nutrients and water have to be applied regularly throughout the plant's growing season.

Watering

The amount of watering needed depends on the time of year, the weather, the growing mix and the type of orchid.

In hot weather, while plants are in active growth, most need daily watering or at least spray misting to maintain moisture around their roots. In conditions where humidity is low and the temperature high, plants may need spraying with water three or four times a day rather than watering around the roots. Damping down the floor of a glasshouse can lower the temperature and increase the humidity when needed.

In cool weather and when plants are completely or fairly dormant, some types are seldom watered. Many originate from regions with well-defined wet and dry seasons. In the wet season they make new leaves and pseudobulbs, in the dry season they rest from growth but produce flowers just before their new growing season begins. In most cases allow them to dry out between waterings. Other types may need watering about once a week during their rest period. Plants constantly watered while they are dormant are likely to collapse and die from root rots. No one can tell you exactly when to water: you have to learn by observation and experience. More orchids are killed through overwatering than anything else. Just remember orchids have growth periods and rest periods.

Feeding

As with water, apply fertilizer only to orchids in active growth. Plants can absorb nutrients only in solution and so regular watering helps orchids absorb the fertilizer. Never apply fertilizer to plants that are bone dry. Always water them first, apply the fertilizer, then water again. It is better to underfeed rather than overfeed: too much can burn and cause problems while giving less simply means that plants grow more slowly.

Garden centres stock fertilizers specially formulated for orchids, some to promote vegetative growth and others to promote flowering, the idea being that you switch from one to the other at a certain stage in the growing season. Other fertilizers are sold as complete plant foods. Soluble plant foods can be applied through the growing season or slow release granular fertilizers applied as new growth begins and again after three or four months. Fertilizers high in nitrogen are best used to promote growth while those high in potassium and phosphorus will help promote flowering. Details of fertilizer ratios are listed on the sides of packets or bottles.

Orchids growing on slabs should be fertilized by spraying with dilute soluble fertilizer. Spray when damp and the weather overcast so that no burning occurs.

PROPAGATION

Many orchids are propagated in nurseries by tissue culture, allowing a huge number of plants to be grown from a small amount of the parent plant. Plants grown from seed are also cultured in flasks. This is useful for sending plants to other countries as the plants have been grown in sterile conditions and will therefore pass quarantine regulations. They are transferred from flasks to individual pots when large enough to survive.

At home, most orchids are propagated by division of existing plants, by removing offsets that have some developed roots or by growing new plants from dormant pseudobulbs. Methods vary and the technique best suited to each group is outlined in the plant entries. Division is best done straight after flowering, as new growth appears.

WHAT CAN GO WRONG?

Orchids are remarkably trouble free. Adequate spacing of plants, good ventilation and good cultural practice should minimize problems. Remove dead or decaying material to keep plants clean and looking good. Isolate any sick plants from the rest and wash your hands and disinfect any tools before handling healthy plants. If you have to use chemicals for pest or disease control, do not spray buds or flowers as they may be distorted. It is usually the newest, most tender parts that are affected.

Pests

- Snails can chew holes in buds and flowers. Remove them by hand or place a few pellets on top of the pots to catch them before they reach the flowers.
- Vine weevils also chew holes in buds and flowers. They are hard to control but some insecticidal dusts may be able to help.
- Several types of scale insect may attack a range of orchids. Small infestations can be gently washed or wiped off with a damp cloth. For heavy infestations spray with an insecticide.
- Red spider mites may be prevalent in warm, dry weather. They can make the foliage mottled and dull looking. Overhead watering and misting the undersides of the leaves helps to discourage them. However, you may actually need to dust or spray the plant with a registered miticide.

Diseases

- Fungal leaf disease. There is a large range of fungal leaf diseases that may attack various orchids, especially in overcrowded or very humid conditions. It may be quite difficult to control them without resorting to a fungicide. Improving ventilation and spacing out the pots should help. Avoid overhead watering and try not to water late in the day if you are plagued with fungal problems.
- Virus disease. There are a number of virus diseases that may attack orchids and there is no cure. Symptoms are variable and may include pale greenish-yellow spots, streaks or patterns of brown, black concentric rings or other patterns along the leaf blade. Serious orchid growers will generally destroy plants affected by virus disease. If you do not want to do this you must isolate the affected plant. Virus diseases may be transmitted through sap-sucking insects such as aphids, which must be controlled, and you must wash your hands and disinfect tools after working on a plant suspected of being diseased.
- Bulb or root rots are caused by organisms found in the potting medium. The organisms flourish when conditions are overwet. Rotted pseudobulbs or roots must be cut away cleanly from the healthy ones and all the old mix washed off. Scrub and disinfect the pot before refilling it with fresh potting mix and replacing the plants. Make sure the mix is very well drained; do not overwater. Some fungicides, used as drenches, can help control the problem.

Mounting orchids on bark

WHY GROW ON BARK?

Many of the orchids grown in cultivation are in nature tree-dwelling, or epiphytic. They use the trees in the rainforests as a perch on which to grow, enabling them to grow nearer to the light. Orchids that grow in this way often have a creeping habit and produce a lot of aerial roots. These are two characteristics that make them difficult to grow in a pot. They are better mounted on a piece of cork bark and allowed to grow across its surface.

PREPARING YOUR PLANT

First you will need to select the plant that you are going to mount. It must have quite a creeping habit of growth with an elongated rhizome between the pseudobulbs. A plant that has a tight cluster of pseudobulbs will not fit well and will have to be regularly remounted. The advantage of growing orchids on bark is that they can remain there for many years without having to be disturbed. If the plant outgrows the first piece, it can be trained on to another piece attached to the original.

The plant should have a healthy active root system to make it easier for it to quickly establish itself in its new position. When moving any plant, wait until it is just starting its new growth as this is when new roots are formed and the plant will suffer the least disturbance.

Choose a piece of cork bark or even a tree branch on which to mount your orchid that will give it enough room to grow for at least a few years. You may even want to mount several plants on the same large piece of wood, to make an interesting feature in your greenhouse.

Other equipment that you will need includes some sphagnum moss and coconut fibre, which will combine to form a moist area around the roots. If you cannot find these particular items then a mixture of similar moist but fibrous substances will do. Some plastic coated wire or fishing line should be used to attach the plant to the bark. After a period of time, once the plant has become established and rooted itself on to the bark, the wire will become obsolete and can be removed if wanted.

It is important to choose a type of plant, preferably with a healthy root system, that will lend itself to being mounted on bark. You will also need plastic-coated wire, pliers, a piece of cork bark and some sphagnum moss or coconut fibre.

MOUNTING THE ORCHID

Take the plant out of its pot, clean away the old compost and trim any dead roots – it will produce new ones once established. Carefully wrap the moss and fibre mixture around the base of the plant, where the roots will weave their way into it, and position the plant on the bark. Secure it with some wire and, while holding it in place, tighten the wire or fasten the fishing line. Take care not to let the wire cut into any part of the plant. It is best to pass the wire in between the pseudobulbs, across the rhizome and away from the new shoots to avoid damage. Attach a wire hook to the top of the bark so you can hang it up in your greenhouse. Make sure it is sprayed or dunked in water daily to prevent drying out.

Remove the orchid from its pot, clean the compost from the roots and trim back a little. Mix moss and fibre together to form a pad on which to place the plant to provide a moist surface for the roots.

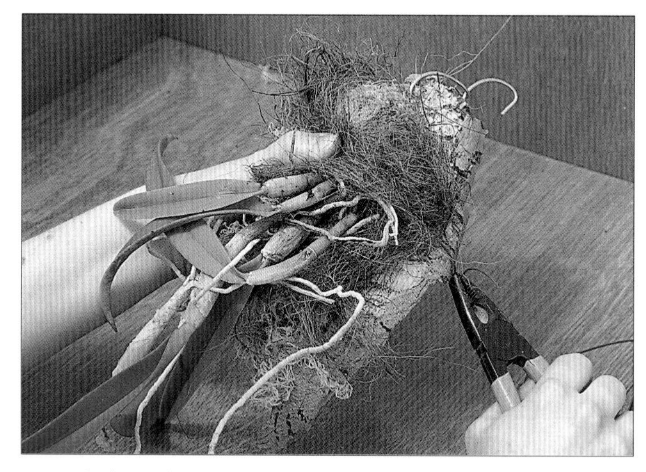

Wrap the base of the plant with the moss mixture and position on the bark. Tie a piece of plastic-coated wire around the base of the plant, avoiding shoots or roots, and tighten just enough to keep the plant in place. Attach a hook and hang where it can be regularly sprayed.

CONVERTING A FISH TANK

AN INDOOR GROWING ENVIRONMENT

A house can be too dry an atmosphere to keep orchids successfully. This is especially true for the small growing ones that tend to dry out more quickly than the plants in larger pots. An interesting and fun way of growing them in the home is to convert an old, disused aquarium into a miniature orchid house. A humid atmosphere will be created inside the glass tank, preventing the plants from becoming too dry. The orchids can be mixed with companion plants that enjoy the same conditions as long as they stay small so the tank is not outgrown quickly.

TOP RIGHT An old, disused fish tank makes an ideal growing environment for miniature orchids. Choose some plants which will stay small and not quickly outgrow the space. Companion plants that will also remain small are helpful to the overall environment.

MIDDLE Making sure that the tank will not leak, fill the base with a layer of expanded clay pellets. These absorb the moisture sprayed on them which then gradually evaporates around the plants creating a humid atmosphere. Add some decorative pieces of wood, bark or rock to create an interesting feature of your indoor garden.

BELOW Lastly, include the finishing touches of the plants including maybe some miniature ferns and foliage plants to complement the orchids. Regularly spray the plants and pellets in the tank to create humidity and remove the plants when you are actually watering them to prevent a build up of too much water in the base. Place near a window and use some shade cloth to cover the tank if too much bright sun is available. A lid is not essential but can be used.

GREENHOUSE CULTIVATION

WHY USE A GREENHOUSE?

To get the best out of your orchids it is advisable to set up a greenhouse especially for them. This means that you can get the growing conditions exactly right. Within this greenhouse you can regulate the temperature and humidity of the air throughout the year and determine how much light, water and ventilation the plants receive. A better environment can be created and so a wider range of orchids can be grown in a greenhouse than in the home.

POINTS TO CONSIDER

You may wish to convert an existing greenhouse or start afresh with a newly built structure. Whatever you decide, it is important that the greenhouse is positioned in the right place. As orchids prefer a shady environment, it is best to position your greenhouse in a shaded part of the garden, near to deciduous trees as these will provide shade in the summer and let in the light in winter when they have lost their leaves. Some extra shade may be needed during summer though, when the sun is at its brightest, as orchid leaves can be easily scorched. Use paint shading on the glass or netting, which can be removed for the winter, or a combination of the two depending on your own greenhouse's situation. The orchids should be kept in dappled sunlight to gain the right amount of light; if too dark then their growth and flowering will be inhibited.

Most traditional greenhouses have glass in their roofs but there are more modern materials available now that need less maintenance, including twin or triple thickness polycarbonate sheeting. This is a rigid plastic sheeting that is very strong and acts as an extremely good insulator, cutting down on the heating requirements for the winter months. It does not matter what type of roofing material you decide to use, but it is important that you make sure that the greenhouse is well ventilated. On hot days, the temperature can rise dramatically and will quickly suffocate the plants inside if there is not enough ventilation available in the greenhouse. Side panels that open in the walls and roof ventilators should be incorporated into your greenhouse so that they can be opened on hot days to give plenty of air movement.

Brighten up an ordinary conservatory with a few orchids. They will enjoy the light, airy environment as well as added moisture from daily misting. Some winter insulation may be necessary as seen here with the bubble polythene covering the door.

PREPARING YOUR GREENHOUSE

Heating is very important for orchids during the colder months of the year. Cooler growing orchids enjoy a drop in temperature but even they will not tolerate temperatures much below 10°C (50°F). Warmer varieties need a few more degrees, a minimum of 15°C (60°F), so need an extra heat supply. This can be supplied by an electric, gas or oil fuelled greenhouse heater. Take the advice of a good supplier to choose the right equipment for your particular set up. A maximum/minimum thermometer is also a very useful piece of equipment as it allows you to keep a check on what the temperature is dropping to at night.

Benches and shelving are ideal ways to arrange your plants at the height that is comfortable for you and your plants. Large plants can be stood lower down, or even on the floor, while smaller pots can be placed on the benches or shelves. If space is at a premium then use the area above the shelves to hang plants up too. This is ideal for orchids that like a bit of extra light and they, in turn, will give a little extra shade to the plants growing below them. If your orchid collection is just beginning then they may have to share their space with other plants already living in your greenhouse. This is fine as long as they all need the same conditions. If the orchids are not compatible with your other plants then it may be necessary to partition the greenhouse with transparent, UV-treated polythene to form two or more separate growing environments. This is also a good idea if you plant to grow a mixed collection of some warm and some cool growing orchids.

CREATING THE RIGHT ATMOSPHERE

One of the best things about growing orchids in a greenhouse is that you can create a humid atmosphere in there for them. Spray the floor of the greenhouse regularly with water, especially in warm weather to keep the temperature down and the air moist. Another way of creating humidity is by growing other types of shade loving plants underneath the benches, such as ferns. This all adds to the overall atmosphere and the orchids will grow better because of it.

Spraying and watering can be done with a watering can, but as your orchid collection grows it may take many trips to fill up the can. A more convenient alternative is to install a hose pipe system that can then reach all parts of the greenhouse, maybe even with a watering lance attached to the end of it. This will allow you to regulate the amount of water that you give to each individual plant; as well as giving the choice of a variety of spraying and misting head attachments.

ABOVE If you have the space then why not create your very own walk-through tropical paradise with impressive foliage plants accompanying the orchids which can live happily in this habitat. Climbing plants can create shade but be careful that they do not harbour pests.

RIGHT Make the best use of your greenhouse by creating maximum bench space and also using the space above the plants to hang some orchids in baskets and hanging pots. Foliage plants placed underneath the benching help with the humidity as well as the use of a humidifier, seen here on the floor.

ANGULOA

Tulip orchid

The Anguloa clowesii *is known as the 'Tulip Orchid' because of its amazing cup-shaped flower which has a rocking lip inside it.*

A strong scent and incredibly waxy blooms are typical of this family as is seen in this Anguloa virginalis.

FEATURES

Terrestrial

This fascinating orchid – known as the tulip orchid due to its tulip-shaped flower – has another common name, cradle orchid. This names describes the lip inside the cup-shaped flower which rocks back and forth when the bloom is tipped. The genus *Anguloa* is closely related to the lycastes with which they can interbreed, making an *Angulocaste*. The species originate from Colombia, Venezuela and Peru and are mainly terrestrial plants. Large, broad leaves are produced in the summer months from the new growth, but in the autumn these die off as the plant goes into its deciduous rest for winter. Its dark green pseudobulbs will then lie dormant until the following spring.

ANGULOA AT A GLANCE

Better for the experienced grower. The flowers last a long time and are hightly scented. Will reach 60cm (24in).

JAN	rest
FEB	rest
MAR	water and feed
APR	water and feed
MAY	flowering, water and feed
JUN	flowering, water and feed
JULY	water and feed
AUG	water and feed
SEPT	water and feed
OCT	rest
NOV	rest
DEC	rest

RECOMMENDED VARIETIES

A. cliftonii (yellow with red markings)

A. clowesii

A. uniflora (white)

A. virginalis

Another feature of this genus is that the long-lasting flowers are strongly scented, often similar to a liniment fragrance.

CONDITIONS

Climate The anguloas need cool conditions with a winter minimum of 10°C (50°F), and summer maximum of 30°C (86°F).

Aspect Due to the soft, annual leaves, shade is required in the summer.

Potting Mix A medium grade bark is ideal with some finer grade bark or peat mixed in.

GROWING METHOD

Propagation It is quite a slow growing orchid, often making just one pseudobulb a year, so will not increase in size enough to divide easily. Back bulbs will sometimes re-grow if removed and potted up separately.

Watering While the plant is in its winter rest, and is leafless, the compost should be kept dry. Watering can be resumed at the start of the new growth in spring. While it has leaves the plant should not dry out. In autumn the new pseudobulb will have been completed and the leaves will turn brown and drop off. Stop watering the plant at this point.

Feeding Plants will benefit from regular feeding while in growth so that the new pseudobulb can develop in the short growing season.

Problems As long as the compost is kept dry during winter then no problems should occur. Avoid water collecting inside new growth.

FLOWERING SEASON

Late spring to early summer.

ANSELLIA

Leopard orchid

The Anselllia africana *is known as the leopard orchid in its native Africa, where it is very widely distributed.*

Copious aerial roots are typical of ansellias. These often point upwards and, once dried, are an excellent defence mechanism.

FEATURES

Epiphytic

This is a very small genus of orchids containing only two species, which are actually regarded as different varieties of the same species by some botanists. *Ansellia africana* is an extremely widespread epiphytic orchid, and is found in many areas of Africa growing prolifically in its natural habitat. A variable orchid, some types have a lot of brown spotting which almost covers the flower; others are plain yellow. This is a large, substantial plant with tall, thin pseudobulbs topped with several pairs of leaves. The flower spike comes from the top of the newest pseudobulb and is long and often branched. This orchid is known for its extensive aerial root growth, producing them over the side of the pot. These roots need spraying regularly to prevent them from drying out and to keep the green growing tips healthy and active.

ANSELLIA AT A GLANCE

Good hot house orchid, it can reach 1m (3¼ft) high. Flowers are 5cm (2in) across on branching spikes, up to 50cm (20in).

JAN	rest
FEB	rest
MAR	rest
APR	rest
MAY	flowering, water and feed
JUN	flowering, water and feed
JULY	flowering, water and feed
AUG	water and feed
SEPT	rest
OCT	rest
NOV	rest
DEC	rest

RECOMMENDED VARIETIES

A. africana

A. gigantea

CONDITIONS

Climate Because of the warm natural habitat that this orchid is used to, it is advisable to keep it in a range of 15–30°C (60–86°F). It grows best when placed in a warm greenhouse or conservatory with good humidity in summer.

Aspect Enjoys good light; bright sun in summer will help in flowering but avoid scorching. Needs maximum light in winter.

Potting Mix Free-draining coarse bark compost.

GROWING METHOD

Propagation Can be divided after some years but will not increase in size very quickly.

Watering It is important to give this orchid a rest during the winter months, letting the compost dry out completely before giving a little water if any shrivelling occurs in the pseudobulbs. The new growth will start again in the spring; gradually increase the watering during the growing season.

Feeding Only add soluble fertilizer to the water when the plant is in active growth during the summer. Feed every two or three waterings when the new pseudobulb is forming.

Problems Problems can occur if the roots are kept too wet during the rest period, so take care to keep dry in winter.

FLOWERING SEASON

Long-lasting throughout the summer.

Aspasia

Aspasia species

This charming, compact plant is easy to grow and will re-bloom very easily, even as a houseplant in a room with a little warmth.

The pretty, star-shaped flowers of Aspasia lunata *nestle around the base of the attractive, leafy plant. Several single flowers are produced.*

FEATURES

Epiphytic

This is a compact and flowering orchid perfect for the beginner. *Aspasias* are a small group of orchids, the genus containing only about ten different species. These originate from the tropical Americas, and are found growing from Nicaragua to Brazil over quite a widespread area. Although not a common orchid, it is actually quite easy to grow, increasing freely in size and producing flowers regularly and easily. Its habit is fairly compact, the height of the soft-leafed pseudobulbs reaching only around 15cm (6in), making it ideal for a small collection. Due to its naturally epiphytic nature, the plant has a creeping habit with an elongated rhizome connecting the pseudobulbs. This means that it quickly outgrows pots and is in need of annual re-potting. However, with orchids that have a tendency to do this, the plant is often happier out of the pot than in it.

ASPASIA AT A GLANCE

Easy to grow. Compact, attractive plant up to 12cm (5in) high. Flowers, often in succession, 3cm (1¼in) across.

JAN	rest
FEB	rest
MAR	flowering, water and feed
APR	flowering, water and feed
MAY	flowering, water and feed
JUN	flowering, water and feed
JULY	flowering, water and feed
AUG	flowering, water and feed
SEPT	water and feed
OCT	rest
NOV	rest
DEC	rest

RECOMMENDED VARIETIES

A. epidendroides (brown petals with a purple and white lip)

A. lunata

CONDITIONS

Climate Slightly cold sensitive so prefers a minimum temperature of 12°C (54°F) in winter, up to 30°C (86°F) in summer.

Aspect Has pale, soft leaves so a little shade in summer will prevent paling or scorching.

Potting Mix A medium grade general bark potting mix.

GROWING METHOD

Propagation This orchid readily produces new growths and so multiplies quite quickly. Therefore, the plant can be divided every few years if required but will do well to be left alone to grow into a specimen plant, which will produce many flowers at once.

Watering The plant does not always follow a strict seasonal pattern so keep it simple by watering more frequently only when in active growth and reducing this to a minimum when not.

Feeding Use a higher nitrogen feed when applying in the growing season, a weak solution every two to three waterings.

Problems If cultural conditions are suitable then it should have no specific problems.

FLOWERING SEASON

Varies but mostly spring and summer. Flower buds emerge from the base of the new growth and stay around plant's base.

BIFRENARIA

Bifrenaria species

Bifrenaria tyrianthina *is an unusual species to grow and is stunning with its large pink blooms on upright stems.*

The more commonly seen B. harrisoniae *has a curiously furry texture and a contrasting purple lip. It is a popular and easy orchid to grow.*

FEATURES

Epiphytic

The charming and popular bifrenarias were once classified as cymbidiums, and show a resemblance to them in their flower shape. These orchids, however, mostly come from Brazil but can also be found widely distributed throughout Panama, Trinidad, northern South America and Peru. *Bifrenaria* has always been a popular orchid for beginners and proves easy to grow and flower in the amateur's cool mixed collection. It is a compact growing plant with long-lasting, heavily textured flowers sitting around the base of the plant. As a bonus the flowers are sweetly scented. They are epiphytic orchids in nature, growing on the higher branches of trees in the South American rain forests. It is possible to grow these orchids quite successfully in a cool to intermediate greenhouse, conservatory or even on a windowsill with other companion plants. If in a greenhouse, the plants could be grown in a hanging basket near the light coming through the roof.

BIFRENARIA AT A GLANCE

Dark green broad foliage reaches 20cm (8in) from pseudobulb. Single flowers with bearded lip low at base.

JAN	rest
FEB	rest
MAR	flowering
APR	flowering, water and feed
MAY	flowering, water and feed
JUN	water and feed
JULY	water and feed
AUG	water and feed
SEPT	rest
OCT	rest
NOV	rest
DEC	rest

RECOMMENDED VARIETIES

B. atropurpurea (dark purple-brown)

B. harrisoniae

B. tyrianthina

CONDITIONS

Climate A temperature range of 10–25°C (50–77°F), with ventilation in the summer months.

Aspect Good light all the year round but provide some shade in the hottest months.

Potting Mix A general medium grade bark compost is ideal with good drainage qualities.

GROWING METHOD

Propagation This orchid is a fairly slow grower so it could be a few years before it is ready to be divided. Best to leave as a specimen plant for as long as possible. May propagate from back bulbs that are removed and are potted up separately at potting time.

Watering Keep compost on the dry side during the winter months when the plant is not growing. With the onset of the new growth in the spring resume watering to get the new pseudobulb plumped up by the autumn.

Feeding The plant responds to a light feeding during the growing season. Use a higher nitrogen plant food every two or three waterings.

Problems No specific problems known if the cultural conditions are suitable.

FLOWERING SEASON

Long-lasting through spring and summer.

Bletilla

Chinese ground orchid

The slightly ridged pattern on the lip of this Chinese ground orchid is very pretty when viewed at close quarters.

Reliable and easy-care, clumps of Bletilla striata *multiply readily. They enjoy filtered sun in most conditions.*

FEATURES

Terrestrial

Bletilla is one of a small group of terrestrial orchids from China, Japan and Taiwan, this is a very easy orchid to cultivate. It is deciduous, dying back to ground level in autumn or early winter. It grows from a pseudobulb that looks like a corm, each producing about three bright green pleated or folded leaves up to 40cm (16in) long. Flowers are slightly bell shaped, cerise-purple to magenta, and carried on one slender stem – there may be 10–12 blooms on a stem. The lip of the flower is beautifully patterned in white and cerise. There is a white form, 'Alba', but it is not as vigorous. The blooms usually last a few weeks if conditions are good.

BLETILLA AT A GLANCE

Easy to grow for cool conditions similar to an alpine house. Bright purple flowers give a good winter show.

JAN	flowering, rest
FEB	flowering, rest, re-pot
MAR	flowering, water and feed, re-pot
APR	flowering, water and feed, re-pot
MAY	water and feed
JUN	water and feed
JULY	water and feed
AUG	water and feed
SEPT	water and feed
OCT	rest
NOV	rest
DEC	rest

RECOMMENDED VARIETIES

B. striata

CONDITIONS

Climate This is a very cool growing orchid and will thrive in a greenhouse if it is kept frost-free, at a minimum temperature of 5°C (41°F). Can be placed outside in summer and even planted in the ground in frost-free areas.

Aspect This orchid needs to be planted in a sheltered spot with dappled shade. It must have protection from hot sun in the middle of the day.

Potting mix Needs well-drained mix or soil with plenty of humus so that it can retain moisture during dry periods.

GROWING METHOD

Propagation Congested clumps of plants can be divided in late winter to early spring. Replant the pseudobulbs at the same depth as they were previously and remove dead leaves.

Watering Needs plenty of regular watering during hot weather through spring and summer. When the leaves begin to yellow off, decrease watering and then stop as the plant dies down.

Feeding Apply slow release fertilizer in spring or give liquid plant food at half strength every two or three weeks during the growing season to help increase growth.

Problems No specific problems are known. Foliage burns if exposed to too much hot sun in summer and plants will rot in a poorly drained medium. Keep the plant dry when not in leaf.

FLOWERING SEASON

Flowers between late winter and spring.

BRASSIA

Spider orchid

The spider orchids are fascinating with their long, thin petals and unusual green colouring, seen in this species, Brassia verrucosa.

This larger flowered hybrid, Brassia *Rex has the characteristic sweet fragrance that these orchids are known for.*

FEATURES

Epiphytic

This group of epiphytic orchids are extremely popular. The attractive leafy plants are easy to keep and flower well when given plenty of light. They are known for their spidery flowers, which give them their common name. The blooms are long lasting, staying on the plant for many weeks and giving off a very pleasant fragrance. There have been many hybrids developed between the species, giving extra size and quality to the flowers, for example *B.* Edvah Loo. Brassias have also been used to make hybrids with other genera such as miltonias and odontoglossums, to produce miltassias and odontobrassias, which inherit the star-shaped flowers and showy appearance of the parent. Grow the brassias in hanging baskets near the light to achieve maximum flowering potential. In this environment they also produce prolific aerial root growth.

BRASSIA AT A GLANCE

Easy to grow. Fragant flowers on long spike, up to 50cm (20in). Compact plant, 15–20cm (6–8in) high.

JAN	rest
FEB	rest
MAR	rest
APR	rest
MAY	flowering, water and feed
JUN	flowering, water and feed
JULY	flowering, water and feed
AUG	flowering, water and feed
SEPT	rest
OCT	rest
NOV	rest
DEC	rest

RECOMMENDED VARIETIES

B. giroudiana
B. maculata
B. verrucosa
B. Edvah Loo
B. New Start
B. Rex
(all green/yellow)

CONDITIONS

Climate They thrive in a cool or intermediate temperature; 10–12°C (50–54°F) at night in winter to 20–25°C (68–77°F) in summer

Aspect Brassias need light to encourage flowering, so place in a south facing aspect with a little shade from the brightest sun.

Potting Mix Medium or coarse mixture of bark chippings.

GROWING METHOD

Propagation The plant should produce several growths from one pseudobulb when it reaches a mature size, which will increase the size of the plant quickly. It can then be divided. Make sure divisions are not too small otherwise they will not flower well. Keep a minimum of four to six pseudobulbs.

Watering Keep the compost moist all the year round. Watering can be reduced in winter months to prevent compost waterlogging. Regular watering while in growth and spraying of leaves and aerial roots will benefit the plant.

Feeding Only apply fertilizer when the plant is in growth. Use a water-soluble feed and pour through the compost as well as adding it to the water used for misting leaves and roots.

Problems Brassias may not flower well if not enough light is provided, especially in winter.

FLOWERING SEASON

Generally late spring and summer.

BULBOPHYLLUM

Bulbophyllum species and hybrids

Bulbophyllum *Jersey has a wonderful rich red colouring and an unusual shaped flower inherited from its parents.*

This is B. lobbii, *an orchid that is well known for its lip that rocks back and forth when the flower is moved.*

FEATURES

Epiphytic

The genus *Bulbophyllum* is one of the largest in the orchid family and includes some of the most extraordinary looking flowers in the orchid kingdom. It is closely related to, and often classified with, the genus *Cirrhopetalum*. They are extremely widespread, being found in South East Asia, Africa, Australia and the tropical Americas. The habit and appearance of the plants and flowers are as variable as their place of origin. Some have tiny flowers that you need a magnifying glass to see; others have large, unusually shaped, showy blooms. A characteristic of many of these orchids is a curiously rocking lip, which attracts certain pollinating insects to the flowers. Some are also fragrant, however this is not always pleasant. The orchids try to attract carrion flies, so they send out the scent of rotting meat. They make good specimen plants, growing well in, and over the edge of, hanging baskets, in which they can stay for years.

BULBOPHYLLUM AT A GLANCE

Strange appearance, with a variety of shapes and sizes, from 3–30cm (1¼–12in) high. Flowers 2mm–8cm (⅛–3in) across.

JAN	rest
FEB	rest
MAR	rest, re-pot
APR	rest, re-pot
MAY	water and feed, re-pot
JUN	water and feed
JULY	water and feed
AUG	water and feed
SEPT	water and feed
OCT	rest
NOV	rest
DEC	rest

RECOMMENDED VARIETIES

- *B. careyanum* 'Fir Cone Orchid'
- *B. graveolans* (cluster of green and red flowers at the base)
- *B. lobbii*
- *B. macranthum* (purple)
- *B. purpureorachis* (brown flowers creeping up a spiral stem)
- *B. vitiense* (small pink)

CONDITIONS

Climate Due to the widespread nature of these orchids, there are both cool and warm growing species available, so check with your supplier when making a purchase. Most of the Asian types are cool, whereas the African species tend to be warmer.

Aspect They can take a lot of light so grow well in a hanging basket near the greenhouse roof.

Potting Mix Need an open medium or coarse grade bark with even some perlite or larger perlag mixed in to make it free draining.

GROWING METHOD

Propagation Most will grow quickly into large clumps with multiple growths so can be divided after only a few years. Alternatively leave growing in a basket for many years until the orchid completely envelops the basket.

Watering Let bulbophyllums dry out in between waterings and take care not to overwater them when in growth. If the pseudobulbs start to shrivel then they are too dry. By growing them in a coarse bark in open baskets they should not stay too wet.

Feeding Give feed only when in growth, and apply this as a foliar feed by spraying it on the leaves as well as pouring through the compost. A weak dilution every two or three waterings is ideal.

Problems No specific problems are known if the cultural conditions are suitable.

FLOWERING SEASON

Depends on the species or hybrid grown.

CALANTHE

Calanthe species and hybrids

Chalk white Calanthe triplicata *flowers in midsummer and is an evergreen type. A succession of blooms makes it long-lasting.*

The intricate flowers of Calanthe vestita var. rubra oculata *appear in arching stems in winter after foliage has died back.*

FEATURES

Terrestrial

This group of mainly terrestrial orchids may be evergreen or deciduous. Originating in south-east Asia, China, Japan, tropical America and the Pacific, they will grow in the ground or pots outdoors in the tropics and subtropics, but need protection elsewhere. Flowers are produced in long sprays and last for many weeks. Leaves are medium to dark green and folded or pleated. Calanthes were among the first exotic orchids cultivated in Britain and Europe. All parts of the plant turn blue if bruised or damaged because of the presence of the dye indican, which produces indigo.

Deciduous *Calanthe vestita*, with white or pale pink flowers and a deep red lip, and C. Veitchii, with rose-pink flowers and a dark lip, shed their foliage before flowering in autumn or winter. This plant is a very old hybrid.

Evergreen *C. triplicata* has broad, pleated leaves growing to 50cm (20in) or more; white flowers appear on a tall stem up to 1m (3¼ft) high, usually in spring or summer. *C. discolor*, with its narrower leaves, bears spring flowers in total contrast to the others mentioned. Flowers are purple-brown or green with the lip either pink or white. This particular plant is among the most cold-tolerant of the species.

CALANTHE AT A GLANCE

A deciduous orchid that blooms when in leafless dormant period with tall sprays of pink/white flowers. Large leaves.

JAN	flowering, rest
FEB	flowering, rest
MAR	flowering, rest
APR	re-pot, water and feed
MAY	re-pot, water and feed
JUN	water and feed
JULY	water and feed
AUG	water and feed
SEPT	water and feed
OCT	rest
NOV	rest
DEC	flowering

RECOMMENDED VARIETIES

- *C. bicolor* (yellow)
- *C. discolor* (brown/ white)
- *C. triplicata* (white)
- *C. vestita* (pink/white)
- C. Grouville (deep red)
- C. Veitchii (white/red)

CONDITIONS

Climate Most calanthes need a minimum temperature of 15°C (60°F) and humidity.

Aspect Easiest to grow in warm conditions such as in a heated greenhouse.

Potting mix Needs coarse, well-drained mix or soil. Use bark with leaf mould, compost or coir peat.

GROWING METHOD

Propagation Propagate in late winter or spring from division of rhizomes or from dormant pseudobulbs separated in pairs.

Watering Keep the plant moist throughout the growing season until leaves are shed or flowering begins. Water occasionally in flower; then keep dry until new growth begins in spring. Lightly mist foliage in summer.

Feeding Use soluble liquid plant food regularly during growth period.

Problems In greenhouses is prone to mealybug and mites. Overwatering or poorly drained mixes rot roots. Make sure that it is kept drier in winter.

FLOWERING SEASON

Mainly autumn, winter and spring. A very few bloom in summer. A succession of flowers make these orchids very long-lasting.

CATTLEYA

Cattleya species and hybrids

The delicately fringed lip of this softly coloured Cattleya *hybrid is its most outstanding feature.*

The colour range of Cattleya *is wide but is predominantly pink, purple, cerise and magenta in many popular cultivars.*

FEATURES

Epiphytic

Known as the 'Queen of Orchids', this plant is spectacular. It is used in floral work, especially corsages and bouquets. Flowers can be as large as 20cm (8in) across, although some species have small blooms. The flowers emerge in a sheath from the leaf, either singly or in sprays carrying numerous blooms. Colours include white, cream, yellow, pink, cerise, purple, magenta and crimson, with contrasting lip colours. Some flowers are perfumed. Cattleyas flower only on new growths. They have one or two rather rigid, leathery leaves.

CATTLEYA AT A GLANCE

Flamboyant, with large, usually fragrant, flowers in spring/autumn. Intermediate growing with good light.

JAN	rest
FEB	rest
MAR	rest, re-pot
APR	water and feed, re-pot
MAY	water and feed, re-pot
JUN	flowering, water and feed
JULY	flowering, water and feed
AUG	flowering, water and feed
SEPT	flowers, water and feed
OCT	rest
NOV	rest
DEC	rest

RECOMMENDED VARIETIES

C. bowringiana (purple)
C. loddigesii (lilac)
C. skinneri (pink)
Laeliocattleya Irene Holguin 'Brown Eyes'

Habit Cattleyas grow from an upright pseudobulb on a thick rhizome. They are suitable for pots and hanging baskets.

Habitat Cattleyas are native to Central and South America, occurring from sea level to mountainous areas. Those from mountain regions prefer a fairly constant temperature of about 20°C (69°F), while those from lower elevations prefer quite hot conditions. Most cattleyas in general cultivation are 'warm' growers but there are a few 'cooler' growers too, liking a drop to 15°C (60°F).

Hybrids This group of orchids has been extensively hybridized and the list of cultivars is enormous. The showy blooms of hybrids and cultivars are often heavily ruffled, on both the petals and the lip, and the flower texture is silky. Apart from hybridizing within the *Cattleya* genus itself, few orchid groups have been so widely used for intergeneric breeding; hybridizing with other genera of orchids such as *Laelia* spp., *Epidendrum* spp., *Sophronitis* spp. and *Brassavola* spp. to produce many attractive hybrids which are so popular with growers.

Species Species are likely to have much smaller flowers, often in larger groups or sprays, but nevertheless in a fascinating range of colours and forms. Few are of interest to florists as their blooms are not as large or showy as those of the hybrids, but species are still of enormous interest to collectors and enthusiasts. Many of the popular species are now being propagated by tissue culture, which has removed the need for collecting wild specimens that have become quite rare.

The pink frilly edges on the overlapping petals make this vibrantly coloured Cattleya *a spectacular flower.*

Large white cattleyas are often chosen for bridal bouquets as they need no additional enhancement.

CONDITIONS

Climate Most like tropical or subtropical conditions with atmospheric humidity in all seasons.

Aspect Cattleyas need very bright light. Full sun, however – except in winter – may burn both the foliage and flowers. Shade should be provided during the brightest days of summer. The plants need shelter from very strong wind but there should always be good air movement around them.

Potting mix Needs very open and free-draining mix or soil. Coarse bark alone or bark mixed with charcoal would make a good mix. Do not re-pot every year. Plants are best left undisturbed for about three years at a time. They are often happy to grow over the side of the pot, making copious aerial roots.

GROWING METHOD

Propagation Increase your stock by dividing existing plants after flowering. This should only be done when pots are overcrowded and plants congested. Make divisions that contain at least three pseudobulbs, cut off any old, dead roots and repot. Pots should not be larger than about 15cm (6in) in diameter.

Watering Must be watered copiously and regularly during the warmer months. Small numbers of plants can be watered by soaking them in a bucket of water. However, plants must dry out to an extent between waterings. In winter water only to keep the pseudobulbs from shrivelling. In hot weather damp down surrounds to create humidity.

Feeding Give liquid fertilizer every two to three weeks during spring and summer as the potting mix has no available organic matter. Don't feed plants at all during winter unless they are growing in a heated greenhouse.

Problems Snails, slugs and weevils can chew holes in buds and flowers. Overwatering, especially in cool weather, and heavy potting mixes can kill plants through root rots.

FLOWERING SEASON

Varies; some time in autumn or spring. Plants in bloom can be placed indoors, in a bright spot away from direct sun and heaters.

Fine striping on the lip of this richly coloured cattleya adds to the decorative effect of a very lovely bloom.

COELOGYNE

Coelogyne species and hybrids (cool growing)

A container of graceful Coelogyne *orchids grows best in a cool conservatory, greenhouse or even in the home.*

This cascade of graceful Coelogyne cristata *is an example of highly decorative orchids at their best.*

FEATURES

Epiphytic

The two most common orchids from this group of over 100 species originate in the lower Himalayas. They are epiphytic and survive down to a few degrees Celsius. The leaves emerge in pairs from angled pseudobulbs, which in large plants spill over the container to form strings. Flowers are lightly fragrant. *Coelogyne cristata*, angel or bridal veil orchid, has long sprays of pure white flowers with golden-yellow markings on the lip. *C. flaccida* is less spectacular but has creamy-beige flowers with a contrasting lip at about the same time. The flowers last for several weeks. Both are easy to grow with other cool-growing coelogynes. Some of the other species need warmer conditions.

COELOGYNE AT A GLANCE

Easy to grow, many are scented with attractive flowers. Species vary from 12–25cm (5¼–10in).

JAN	rest
FEB	flowering, rest
MAR	flowering, water and feed, re-pot
APR	water and feed
MAY	water and feed
JUN	water and feed
JULY	water and feed
AUG	water and feed
SEPT	rest
OCT	rest
NOV	rest
DEC	rest

RECOMMENDED VARIETIES

C. cristata
C. flaccida
C. mooreana
C. ochracea

CONDITIONS

Climate Grows easily in cool but humid conditions in a greenhouse.

Aspect Needs very bright light but if grown in too much sun the foliage may become yellowish. 50 per cent shade suits the plant well. If not enough light is given then the plant might not flower to its full potential.

Potting mix Needs very free-draining mix or soil. Coarse bark mixed with coir peat or leaf mould is ideal. Pseudobulbs must be above the surface of the mix.

GROWING METHOD

Propagation Backbulbs can be detached and potted up separately in early spring. Otherwise cut off hanging bulbs that have one or two aerial roots. Plants should be divided only when extremely congested.

Watering Give the plants plenty of water during spring and summer but allow them to dry out between waterings in winter. Some growers like to keep them quite dry in winter. This is sensible in particularly cold conditions as becoming too wet can cause rotting.

Feeding Apply weak liquid fertilizer or granular slow release fertilizer every two or three weeks during active growth.

Problems There are no specific pest or disease problems known.

FLOWERING SEASON

Both of the common species flower towards late winter or spring. Their flowers last for several weeks and make a glorious show if they are brought indoors.

COELOGYNE

Coelogyne species and hybrids (warm growing)

A warmer growing coelogyne is the species C. massangeana, *with its incredibly long, pendant flower spikes.*

Coelogyne *Green Dragon 'Chelsea' AM/RHS is one of the few hybrids and is stunning with its large flowers on long, hanging stems.*

FEATURES

Epiphytic

As well as the wide range of cool growing coleogynes there are also warmer growing types. These are typified by the species *C. massangeana* that originates from Malaysia. As with their cooler cousins, they grow well in pots or baskets, and this species especially needs to be hung up when in bloom due to its pendant flower spike. This can reach up to 50cm (1¾ft) in length and a mature specimen will flower several times in the summer, producing a succession of spikes one after the other. Its large, oval leaves and pseudobulbs keep the plant looking attractive even when the plant is out of flower. Only a few hybrids have been made within this group. The most stunning is C. Green Dragon 'Chelsea' AM/RHS, which has a 60cm (24in) long spike with large green blooms and contrasting black lip.

COELOGYNE AT A GLANCE

Very popular and easy to grow. Can reach 50cm (1¾ft) in length and flowers more than once during summer.

JAN	rest, water less
FEB	rest, water less
MAR	flowering, water
APR	flowering, water
MAY	flowering, water
JUN	flowering, water
JULY	flowering, water
AUG	flowering, water
SEPT	rest, water less, re-pot
OCT	rest, water less
NOV	rest, water less
DEC	rest, water less

RECOMMENDED VARIETIES

- C. *dayana* (yellow/ brown)
- C. *massangeana*
- C. *pandurata* (green)
- C. *rochussenii* (small yellow pendant flowers)
- C. *swaniana* (beige with brown lip)
- C. Green Dragon 'Chelsea' AM/RHS (green/black)

CONDITIONS

Climate Needs warmer conditions when growing. A minimum of 15°C (60°F) in winter and a summer maximum of 28°C (83°F).

Aspect They can take light but will easily burn if too much direct sun reaches the leaves in summer. But if the foliage becomes dark green they may be in too dark a position.

Potting Mix Free-draining coarse or medium grade bark is ideal. Mix in some perlite or perlag if you feel you need extra drainage.

GROWING METHOD

Propagation Will divide after a few years if left undisturbed, and can become quite large if left to grow for many years.

Watering Water and spray the foliage frequently during the summer to plump the new growths up into the large, fat pseudobulbs that are characteristic of these types.

Feeding Add feed to the water while the plant is growing to help the pseudobulbs.

Problems Spray the foliage, including the undersides of the leaves, with water frequently in warm weather, to prevent red-spider mite attacking the leaves.

FLOWERING SEASON

Mostly summer flowering. The blooms only last a couple of weeks but the succession of spikes on a large plant spreads out the flowering time to last several months.

CYMBIDIUM

Cymbidium species and hybrids

Masses of small blooms clustered on a flower stem are a feature of many of the miniature cymbidium hybrids.

Potted cymbidiums are at home outdoors in the summer. Plants in flower should be displayed indoors in a cool position.

FEATURES

Terrestrial

Epiphytic

Cymbidiums are probably the most widely cultivated of all orchids. They originate in temperate or tropical parts of north-western India, China, Japan, through south-east Asia and Australia. There are now thousands of cultivated varieties. These hybrids have flowers classed as standard size (10–15cm [4–6in] across), miniature (about 5cm [2in]across) or intermediate. The leaves are strap-like, upright or pendulous, and 50cm (20in) or more in length in standard growers. The foliage of the miniature plants is narrower and shorter, in keeping with the overall dimensions of the plant. Cymbidiums have a wide appeal as the flowers are decorative and long lasting.

CYMBIDIUM AT A GLANCE

Popular and widely grown. Flowers vary from 4–10cm (1½–4in). Easy to grow in light, cool conditions.

JAN	rest, flowering
FEB	rest, flowering, repot
MAR	flowering, water and feed, re-pot
APR	flowering, water and feed, re-pot
MAY	water and feed
JUN	water and feed
JULY	water and feed
AUG	water and feed
SEPT	rest, flowering, water and feed
OCT	rest, flowering
NOV	rest, flowering
DEC	rest, flowering

RECOMMENDED VARIETIES

C. erythrostylum (white)
C. lowianum (green)
C. traceyanum (brown)
C. Amesbury (green)
C. Bouley Bay (yellow)
C. Gymer (yellow/red)
C. Ivy Fung (red)
C. Pontac (burgundy)

Flowers The range of flower colours covers every shade and tone of white and cream, yellow and orange, pink and red, brown and green, all with patterned or contrasting colours on the lip. Flowers are carried on quite sturdy stems standing well clear of the foliage. As many are in bloom in winter they can give a special lift to the season. Many have flowers that will last six to eight weeks.

Choosing Choose plants both by flower colour and time of blooming. With careful selection you can have a *Cymbidium* in bloom every month from early autumn through to late spring. Selecting plants in flower will tell you exactly what you are getting. Some orchid nurseries sell tissue cultured mericlones of these orchids as young plants: these are much cheaper but you will have to wait three to four years for them to reach flowering size. Backbulbs, if available, are also an option although these too take up to four years to flower. Buying seedlings is also cheaper and you will have the thrill of their first flowering, not knowing in advance what the flowers will be like. The parent plants may be displayed or the nursery should be able to give you an idea of the likely colour range, which extends through the spectrum.

CONDITIONS

Climate Ideal conditions are humid year round, with winter temperatures not reaching much below 8–10°C (46–50°F) and summer temperatures that are generally below 30°C (86°F). It is advisable to place the plants outside in summer. To initiate flowering many, but not all, require a distinct drop between their day and night temperatures.

Pink-flushed white flowers are popular with many Cymbidium *growers. Note the lovely spotting on the lip.*

Aspect These plants will not flower if they do not get sufficient light. They will grow in the open with dappled shade from trees or with only morning sun in summer, but they can actually take full sun almost all day during the winter. A shadehouse with 50 per cent shade is suitable in summer. If you can see that your plants have very dark green leaves then they aren't getting enough light and are very unlikely to flower. These plants need good ventilation and protection from strong winds and rain when placed outside. In winter, place them in the lightest position available within a cool greenhouse, conservatory or in an unheated, light room.

Potting mix Mix or soil must be very free draining. Many species in their habitat grow in hollow branches of trees, in decayed bark and leaf litter. Hybrids can be grown in aged, medium-grade pine bark, or in pine bark plus coir peat, bracken fibre, charcoal or even pieces of foam. Plants need to be anchored and supported but must have free drainage and good aeration around the roots. Prepared, special orchid composts are satisfactory, especially if you only have a few plants. Cymbidiums must always be potted with the base of the pseudobulb either at, or preferably just above, the level of the compost.

Containers Make sure that there are enough drain holes in the container and, if not, punch in more. Cymbidiums are quite vigorous growers and if placed in 20cm (8in) pots they will probably need potting and maybe dividing every two or three years. It is best to pot plants into containers that will just comfortably accommodate their roots. They can then be potted on into the next size when necessary. Plants that have filled their containers and need dividing are best left until spring, and then divided into sections with no fewer than three pseudobulbs per division. If divided soon after flowering in spring the plants will then have a full six months growing season ahead of them to settle into their new pot.

GROWING METHOD

Propagation Grow new plants from backbulbs (older, leafless bulbs). These may look dead but will regrow if they are detached from the younger growths when you are dividing plants after flowering. Clean old leaf bases and trim off any old roots. Plant the bulbs into small individual or large communal pots, about one-third of their depth into a mixture of coir peat and bark. Keep damp but not wet. Once good leaf and root growth are evident, pot up into normal mix. Plants grown this way generally flower after about three years.

Watering Frequency of watering is determined by the time of year. The mix should be moist but not wet. Always give enough at one watering for water to pour through the mix. Plants may need watering daily in summer or only every few days if conditions are wet. Water about once every one to two weeks in winter.

Feeding Feeding can be as easy or as complicated as you like. You can simply give slow release granular fertilizers during the growing season. Or, alternatively, you can use soluble liquid feeds regularly through spring and summer. Some growers like to use high nitrogen fertilizers during spring, switching to special orchid foods or complete fertilizers that are high in phosphorus and potassium in summer. Some of these are applied monthly, others in half strength more often. It is important to follow label recommendations and not to overdo it.

Problems Unfortunately cymbidiums are prone to some diseases, quite apart from the normal range of pests. Virus disease can be a problem and so maintain strict hygiene by disinfecting hands and tools to avoid it spreading. Bulb rots can also be a problem if potting mixes contain too much fine material, which impedes drainage. Fungal leaf spots can occur in crowded or very wet conditions. Keep an eye out for snails, slugs and aphids as your plants come into bud because they can quickly ruin your long-awaited flowers. They will attack both buds and long spikes.

Support Light cane or metal stakes should be used to support the flower spikes as they develop. Carefully insert the support beside the developing spike and use plastic coated tie-wire or string to tie the spike as often as needed to train it into position.

FLOWERING SEASON

In bloom through autumn, winter and spring with just a few summer varieties. Plants in flower can be brought into the house for decoration. Moving a plant in bud from a cool greenhouse to a warm room can make the buds drop so wait until the flowers are open and set until you move it. The cooler the plant is kept, the longer the flowers last, which is on average six to eight weeks.

Cut flowers Professional growers cut the flower spikes a week after all flowers on the spike are open to prevent any check in the plant's growth.

5.
4.
1.

3.

Cymbidiums

1 Cymbidium *Red Beauty* x *Gorey is a standard variety of the* Cymbidium, *which can reach 1m (3¼ft) in height, with tall sprays of rich pinkish red flowers.*

2 A compact type of Cymbidium, *this beautifully coloured* C. *Mini Dream 'Gold Sovereign', has an unusual shade of butter yellow with yellow markings on the lip too, making a striking combination.*

3 Large flowered standard variety Cymbidium *Sleeping Nymph 'Perfection' is sought after for its striking combination of apple-green petals and sepals and yellow marked lip, which is lacking the usual red pigment.*

4 These orchids can reach a fair size if they are left undivided and are of the larger growing type, as this C. *Havre des Pas shows. They will produce a better show if they are allowed to grow into a larger specimen.*

5 For the more modest space available, a compact variety such as C. *Red Valley 'Brilliant' will give a marvellous show, while taking up less space.*

2.

DENDROBIUM

Dendrobium species and hybrids (Asian)

A dark blotch of colour in the throat is a feature of many types of dendrobium, including this Dendrobium nobile.

One of the finest yellow orchids is Dendrobium fimbriatum. *The beautiful, finely fringed lip is greatly admired.*

FEATURES

Epiphytic

By far the largest number of *Dendrobium* species come from sub-tropical and warm regions of Burma, the Himalayas, Thailand, China and Malaysia. Some of the most commonly cultivated are the varieties of species such as *D. nobile*. These are known as soft-cane dendrobiums. Many Asian species have long, cane-like growth which can grow up to 1m (3¼ft) tall, although many others are within the 30–45 cm (12–18in) range. Some are very upright while others have pendulous growth so they must be grown in hanging baskets. The species described here can be grown in a cool glasshouse where night temperatures do not fall much below 10°C (50°F).

DENDROBIUM AT A GLANCE

Popular as houseplants and flowers as cut blooms. Hybrid varieties good for beginners. Various sizes and types.

JAN	rest
FEB	rest, flowering
MAR	water and feed, flowering
APR	water and feed, flowering
MAY	water and feed
JUN	water and feed
JULY	water and feed
SEPT	water and feed
OCT	rest
NOV	rest
DEC	rest

RECOMMENDED VARIETIES

- *D. aphyllum* (pink/cream)
- *D. chrysanthum* (yellow)
- *D. densiflorum* (golden)
- *D. fimbriatum* (yellow, dark centre)
- *D. nobile* (dark pink)
- *D.* Christmas Chimes (white, dark centre)
- *D.* Red Comet (dark pink)
- *D.* Stardust (pink/white)

Types Some species of *Dendrobium* are evergreen while others are deciduous. The latter lose their leaves during their dormant period, which coincides with their dry season. Both types are epiphytic and are found growing on branches of trees or sometimes on mossy rocks in their habitat. The plants will easily grow into large clumps over years, producing a very spectacular show.

Flowers A few dendrobium varieties produce single flowers but most produce large, showy sprays containing numerous flowers. The colour range is vast. White, cream, yellow, pale green, pink, red, maroon, purple and magenta are all represented in this colourful group. Some have flowers of one single tone while many have contrasting blotches of colour in the throat or on the lip of the flower. Many are strongly fragrant.

D. nobile *D. nobile* is a soft-cane stemmed type that can grow from 30–75cm (12–30in) high. The species is pink with deeper cerise tips on the petals and a dark maroon blotch on the lip. The numerous cultivars of this species include many with similar tonings of lavender, purple and red, but some have pure white petals with yellow or dark red markings on the lip.

D. chrysanthum This is an evergreen orchid with canes that often grow over 1m (3¼ft) long. It has a pendulous habit and so is best grown where its stems can hang naturally. Simulating its natural growth this way seems to promote more consistent flowering. Flowers are deep golden yellow with deep red blotches in the lip on a graceful, arching stem.

D. aphyllum *D. aphyllum* (syn. *D. pierardii*) is a deciduous species that prefers to grow in a hanging basket. Its canes can grow to over

The deep gold throat of this softly coloured pink and white hybrid of Dendrobium nobile *provides an exciting contrast.*

In cool, humid conditions Dendrobium nobile *and its cultivars will produce a profusion of flowers.*

1m (3¼ft) and its delicate, pale flowers can best be enjoyed at eye height. The flowers are pale mauve to pink with the palest creamy yellow lip.

D. fimbriatum Another yellow-flowered species of *Dendrobium* that grows with tall, upright canes sometimes reaching over 1m (3¼ft). It is evergreen and the flowers are produced on the tops of canes one year or more old. Flowers may appear even on older canes that no longer bear leaves. This flower is golden yellow and the lip is delicately fringed. The variety *oculatum* is a richer, deeper gold with a deep maroon blotch in the centre of the lip. These dendrobiums can be grown either in heavy pots that have pebbles added in the base to balance the top weight of the canes, or in a hanging basket.

CONDITIONS

Climate The preferred conditions depend on the species. Cool types grow in glasshouses or conservatories, whereas warm types will live on a windowsill indoors.

Aspect These dendrobiums tolerate partial shade to full sun depending on the species. Those with red, bright pink and yellow flowers tolerate much more sun than those with white or pale green flowers.

Potting mix Free-draining mixes must always be used. These may contain coarse bark, tree-fern fibre, sphagnum moss, perlite and even pebbles if extra weight is needed to stabilise the containers. These plants should never be overpotted. Use a container that will comfortably hold the plant roots with a little room to spare.

GROWING METHOD

Propagation All grow from divisions of the existing plants once they have filled their pots. Divide after flowering. Some species produce offsets or aerial growths which can be removed from the parent plant once roots are well developed. Older stems of deciduous species containing dormant buds can be laid on damp sphagnum moss and kept moist until roots develop. This may take several months.

Watering During active growth in warm weather mist or water regularly, two or three times a week. Give only occasional watering in winter; keep those from monsoonal areas dry at this time.

Feeding Feed only during the growing season and not during autumn or winter. Use regular applications of soluble orchid fertilizer.

Pruning Restrict pruning to removal of spent flowering stems. Do not cut out old canes of species such as *D. fimbriatum* which flower on older stems unless they have shrivelled, turned brown or died off.

Problems Chewing and grazing insects, such as snails, slugs, caterpillars and weevils, can all damage these plants but they are a particular nuisance on the flowers. Plants grown in glasshouses may be troubled by mealybugs and mites, as well as fungal diseases if there is poor ventilation.

FLOWERING SEASON

Most flower in spring but the range may be from late winter to early summer depending on growing conditions and the species.

DENDROBIUM

Dendrobium species and hybrids (Australian)

This pretty cultivar of Dendrobium speciosum *has been given the name 'Aussie Sunshine'. Individual flowers are finely shaped.*

Lighting up this garden in late winter to spring are the long cream to yellow trusses of Dendrobium speciosum.

FEATURES

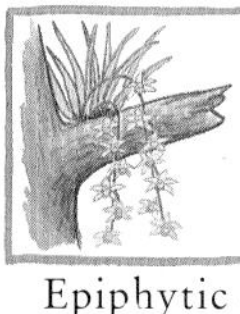

Epiphytic

There are about seventy Australian species of *Dendrobium* but only a few are cultivated outside Australasia. These epiphytic and lithophytic orchids can have cane-like or thick swollen pseudobulbs, but sometimes the pseudobulbs are not visible at all, as in *D. linguiforme*, which creeps over rocks producing small, fleshy, ribbed foliage. Most of these orchids have rather leathery, sometimes very rigid, leaves, and plants not in flower excite little interest. These dendrobiums are extensively grown by amateurs and professionals alike and most people recognise the 'rock lily', *D. speciosum*.

DENDROBIUM AT A GLANCE

Normally three years old before flowering. Flowers last up to six weeks. Strap-like leaves vary from green to silver.

JAN	grow on, reduce watering
FEB	grow on, reduce watering
MAR	re-pot, feed
APR	remove and pot on offsets
MAY	remove offsets, mist foliage
JUN	flowering, mist and water
JULY	flowering, mist and water
AUG	reduce watering
SEPT	reduce watering
OCT	flowering; keep frost free
NOV	flowering, reduce watering
DEC	flowering, reduce watering

RECOMMENDED VARIETIES

D. kingianum (pink)
D. speciosum (yellow)
D. Delicatum (white)

Flowers The majority of the Australian *Dendrobium* species have small individual flowers, although these may be clustered on long sprays. An exception to this is the Cooktown orchid, *D. bigibbum*, floral emblem of the state of Queensland, which has larger flowers in rosy pink to purple.

Cut flowers Most Australian species do not make good cut flowers and blooms may last only three weeks or so on the plant. The flowers of *D. bigibbum*, however, are long lasting and cut well. They are often included in mixed bunches sold as Singapore orchids.

D. speciosum The rock lily or king orchid, *D. speciosum*, is possibly the species most often grown. It has very thick, fairly long, slightly curved pseudobulbs and large, stiff, dark green leaves. It grows on rocks, logs or in hanging baskets, and clumps may spread in time to over 1m (3¼ft) across. In its habitat it sometimes forms large clumps high up in trees. It is very easy to grow and its long, arching sprays of cream to yellow flowers have a light honey scent. Flowering, which usually occurs from late winter through to early spring, can vary from year to year.

D. kingianum The pink rock orchid, *D. kingianum*, is the species of *Dendrobium* most extensively hybridized. Numerous named cultivars are available from specialist growers and some amazing colours are being produced. The true species has short, thickish pseudobulbs topped with leathery leaves and produces little rounded, pink or mauve flowers. There are also forms that have white or almost purple flowers. This is another easy-care orchid which is very appealing.

The Cooktown orchid, Dendrobium bigibbum, *is at home in the tropics, thriving in constant warmth and humidity.*

Rosy purple is one of the naturally occurring colour variations in the species Dendrobium bigibbum.

D. Delicatum Probably one of the most attractive of all these orchids is the hybrid *D.* Delicatum which is a cross between *D. kingianum* and *D. speciosum*. It has long, slender pseudobulbs and produces an abundant display of upright flower spikes that may be white or palest pink, sometimes with a darker lip.

D. gracilicaule A vigorous grower often found naturally on trees that don't shed their bark. It has long, narrow, cane-like pseudobulbs and produces cream to golden yellow flowers in early spring. This is a good orchid to establish in a basket. A natural hybrid of this species and *D. speciosum* is D. *x gracillimum*, which is among the most prolific and free flowering of all these orchids, producing masses of creamy flowers each year.

D. falcorostrum Not so easy to cultivate but well worth the effort is the beech orchid, *D. falcorostrum*, which is becoming very scarce in the wild. Its preferred natural host is the Antarctic beech, *Nothofagus moorei*, which has been overcleared. This beautiful orchid has creamy white, scented flowers in short sprays that develop on top of thickish pseudobulbs that may be 15–25cm (6–10in) long.

D. linguiforme The pseudobulbs are not visible in the tongue orchid, *D. linguiforme*, which creeps over rocks. It produces small, fleshy, ribbed foliage and abundant sprays of feathery, cream flowers appear in spring.

CONDITIONS

Climate Preferred climate depends on the species. Some grow best in the warm, others prefer cool to intermediate conditions.

Aspect Most prefer dappled sunlight, although some tolerate full sun.

Potting mix Many dendrobiums are best grown on slabs, logs, old stumps or rocks. In containers the mix should be very coarse bark, crushed rock and tree-fern fibre.

GROWING METHOD

Propagation Most are best propagated by dividing clumps straight after flowering. Those that produce offsets from the tops of canes can have the offsets gently detached and replanted once they have developed a good root system of their own.

Watering Most of these dendrobiums prefer regular, abundant watering during spring and summer. In the cooler months water very occasionally. Dendrobiums from tropical areas with defined wet and dry seasons should be kept quite dry in their dormant stage to prevent early growth starting.

Feeding Feed with complete soluble fertilizers during spring to early summer or with slow release granules or water-soluble feed. Do not feed at all during autumn or winter.

Pruning Restrict pruning to the removal of spent flower stems.

Problems Most of these plants are fairly trouble-free, although damage can be caused by slugs, snails, caterpillars and weevils chewing flowers and new leaves. Fungal leaf diseases may attack plants that are grown in conditions where humidity is high. This is a real problem in glasshouses, especially if the ventilation is poor.

FLOWERING SEASON

Flowering time depends on species and regional growing conditions. Most flower in late winter to spring.

Disa

Disa species and hybrids

The pale flower of this lovely Disa *cultivar clearly shows the nectar-bearing spur. Red-flowered cultivars dominate most collections.*

Buds and flowers at various stages of development ensure a long flowering season for this Disa *hybrid.*

FEATURES

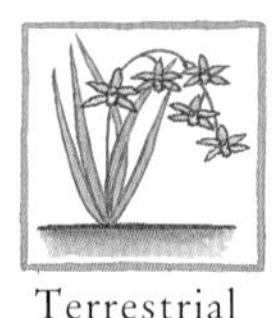

Terrestrial

Only one species of *Disa* is common in cultivation: *D. uniflora*, which has generally bright scarlet flowers. Flowers are mainly scarlet with red and gold venation but some golden yellow ones are found in their natural habitat. This terrestrial genus can be difficult to cultivate successfully as many species come from habitats that have soil that is permanently damp but never waterlogged. These conditions are not very easy for the orchid grower to duplicate. Disas produce a rosette of basal leaves from which a flowering stem over 60cm (24in) high will emerge. Flowers are large, 8–12cm (3–5in) across and borne in groups of mostly three or more blooms.

DISA AT A GLANCE

Unusual and challenging orchid preferring moist conditions. Bright flowers are 5–8cm (2–3in) across.

JAN	water
FEB	water
MAR	water
APR	flowering, water
MAY	flowering, water
JUN	flowering,water and feed
JULY	flowering,water and feed
AUG	flowering,water and feed
SEPT	water
OCT	water
NOV	water
DEC	water

RECOMMENDED VARIETIES

D. uniflora

D. Inca Princess

Origin This group of orchids is mainly native to tropical and southern Africa and is also found in Madagascar.

CONDITIONS

Climate *Disa* needs a frost-free climate. Most are cool growers and tolerate temperatures down to about 5°C (41°F); 25°C (77°F) is the preferred upper limit.

Aspect Grows best in partial shade with some sun early in the morning. In total shade plants will grow but not flower.

Potting mix The mix or soil should be moisture retentive but never soggy. A suitable mix might contain perlite, coconut fibre peat, chopped sphagnum moss and medium-fine bark.

GROWING METHOD

Propagation Divide plants when re-potting. This should be about every two years after flowering, never more than three years apart.

Watering Avoid watering the foliage if possible. Allow water to soak up from below by standing pots in a container of water. Never allow the pots to dry out but greatly decrease water in cool weather.

Feeding Apply soluble liquid fertilizer, at a quarter to half the recommended dilution rate, while plants are actively growing.

Problems No specific problems are known if cultural conditions are met.

FLOWERING SEASON

Generally from early summer to autumn, but this can vary.

Encyclia
Encyclia species

Commonly known as the cockleshell orchid, the shape of the lip that gives Encyclia cochleata *its name can be seen clearly here.*

The scarlet encyclia is E. vitellina, *well known and sought after for its vibrant orange colouring.*

FEATURES

Epiphytic

Another very easy and popular group of orchids. The genus of *Encyclia* was originally classified with the epidendrums so are closely related and require very similar growing conditions. Most of the encyclias are found in Central and South America, but some inhabit Florida and the West Indies, as in the case of *E. cochleata*. This orchid was the first tropical epiphytic orchid to flower in the United Kingdom. Originally discovered in the Americas, it was brought back by the plant hunters to the Royal Botanic Gardens at Kew in London. It has the common name of cockleshell orchid, due to the shape of its upturned, almost black, lip, and is the national flower of the Central American country of Belize. The orchid will flower for months on large specimen plants. Most of the *Encyclia* species are green in colour but *E. vitellina* is bright orange. Many of the green-flowered species are also sweetly scented.

ENCYCLIA AT A GLANCE

Very easy to grow; ideal in most situations. Compact plant, 20–25cm (8–10in) high, flowers 3–6cm (1¼–2¼in) across.

JAN	rest
FEB	rest
MAR	rest
APR	flowering, water and feed
MAY	flowering, water and feed
JUN	flowering, water and feed
JULY	flowering, water and feed
AUG	flowering, water and feed
SEPT	water and feed
OCT	rest
NOV	rest
DEC	rest

RECOMMENDED VARIETIES

E. cochleata
E. lancifolia (cream)
E. mariae (green/white)
E. pentotis (cream/green)
E. radiata (pale green)
E. vitellina (orange)

CONDITIONS

Climate The most popular are cool growing needing a temperature range of 10–25°C (50–77°F).

Aspect A light position is required, out of direct sunshine in summer to avoid scorching. Grow well in hanging baskets in a shaded greenhouse, or as windowsill plants.

Potting Mix An open and free-draining bark mix is good; aerial roots are often made outside the pot.

GROWING METHOD

Propagation Can be divided and propagated after growing into a large plant over a number of years. To ensure flowering the next year, do not make individual plants too small.

Watering Frequently water during summer – its main growing season. When not in growth it should be partially rested; just a little water is required to keep the pseudobulbs plump.

Feeding Feed only when the plant is in active growth when it can gain full benefit from the added nutrients. Apply both in the water poured into the pot and also in the spray given to the leaves and aerial roots.

Problems No specific problems are known if cultural conditions are suitable.

FLOWERING SEASON

Can vary but mainly the summer months.

EPIDENDRUM

Crucifix orchid

A bright pink cultivar of Epidendrum ibaguense *flowers well each year but is not as prolific as the orange-flowered species.*

Tough, reliable and flowering over many months, the crucifix orchid makes a fine garden plant in tropical countries.

FEATURES

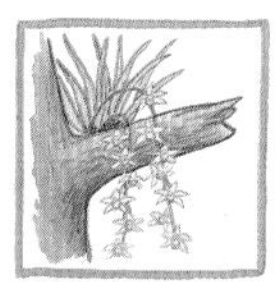

Epiphytic

Terrestrial

This genus and its species is a large and varied group that originated in tropical America. Growth can be rampant and needs thinning out every couple of years. Plants in pots may need support. This orchid will grow without almost any attention but rewards good culture. The species *E. ibaguense* flower is orange and yellow but there are many cultivars with flowers in shades of red, pink, mauve, yellow or white. Individual flowers are fairly small but they are carried in groups on top of reedy canes that can be 2m (6½ft)or more tall. The fleshy leaves are leathery in texture and yellowish green, especially in full sun. *E. ibaguense* is one of the most familiar of the *Epidendrum* species, also known as *E. radicans* or its common name of crucifix orchid, so called because of the cross shape of the lip, when turned upside down.

Habit The crucifix orchid has a tall, reed-like habit of growth so requires some space to grow to its full potential.

CONDITIONS

Climate Grows best in an intermediate greenhouse, with a minimum of 12°C (54°F) in winter.

Aspect Tolerates full sun to dappled shade. Where this orchid is grown in more tropical climates, as a bedding plant in garden borders, it is very tolerant of changing conditions and rough treatment.

Potting mix The mix or soil must be able to drain rapidly but any mixture of coarse bark, crushed rock, gravel, compost or commercial potting mix is suitable.

GROWING METHOD

Propagation Is very easy to propagate as it produces plantlets with aerial roots on the older canes. Detach them and pot up when they are sufficiently developed during the warmer months. Large clumps can also be divided.

Watering Thrives if given regular watering during the warm months of the year and less frequent waterings in winter.

Feeding Give soluble liquid fertilizer every three or four weeks during warm weather or dress the roots with aged cow manure. In warm conditions feed year round.

Problems Trouble-free but fungal leaf diseases can occur if the weather is too cool and wet. Succulent new leaves are often eaten by slugs.

EPIDENDRUM AT A GLANCE

Easy to grow. Repeat flowers all year round. Can reach 1.5m (5ft) in height but flowers are only 3cm (1¼ in) across.

JAN	rest, little water
FEB	rest, little water
MAR	increase water, re-pot
APR	water and feed
MAY	water and feed
JUN	water and feed
JULY	water and feed
AUG	water and feed
SEPT	water and feed
OCT	reduce water
NOV	reduce water
DEC	rest, little water

RECOMMENDED VARIETIES

- *E. cristatum* (brown spotted)
- *E. ibaguense* (red/yellow)
- *E. ilense* (pale cream)
- *E. pseudepidendrum* (green/orange)
- *E. wallissii* (lilac/brown)
- *E.* Pink Cascade (pink)
- *E.* Plastic Doll (green/ yellow)

The hybrid E. *Plastic Doll flowers easily on a very young plant and continues to bloom prolifically over the years.*

Epidendrum *Pink Cascade is almost never out of flower; don't cut off the old flower stem as this will produce buds over and over again.*

FLOWERING SEASON

Can flower almost all the year round.

OTHER EPIDENDRUMS TO GROW

This is a very large group of orchids with many varied species and hybrids available. They are easy to keep and are very rewarding. The plants have long flowering seasons, and some are sequential flowerers, producing more and more buds time after time throughout the year. All need similar cultural requirements to those of the crucifix orchid.

E. ilense This is a very tall growing species from the Tropical Americas, reaching up to 2m (6½ft) in height. The leafy cane-like pseudobulbs are semi-deciduous, losing their leaves after a year or so. The curious flowers come in bunches from the top of the newest and oldest pseudobulb, the old ones re-flowering for many years. The blooms are creamy yellow and around 2cm (¾in) across; the lip, which hangs down, is strangely frilled into lots of tiny threads giving a bearded effect. Flowers can appear on very young plants only 15cm (6in) high.

E. pseudepidendrum Perhaps a confusing name but this is yet another different *Epidendrum*. This species will grow with a similar habit to the *E. ilense* and they grow well when positioned together. This one has a very hard, waxy flower with bright green petals and sepals, swept back from the brilliant orange lip, making a striking combination.

***E.* Plastic Doll** This is a primary hybrid between the two species mentioned previously and it inherits qualities from both. It flowers on very young plants, and has a waxy yellow flower with a frilled lip. When larger, the plant should re-flower continually to give a perpetual show. The plant can grow tall in time as well but generally it is a very easy and rewarding orchid to keep.

***E.* Pink Cascade** Another primary hybrid of *E. ilense* but this time it has been crossed with another species, *E. revolutum*. The Pink Cascade orchid tends to hold more flowers on the stem at one time. As the name suggests, its flowers are bright pink in colour. Do not cut the flower stems on this orchid, as they will flower again and again from the top of the leafy cane.

Epidendrum cristatum *grows very tall and in a large clump. Each stem produces a head of highly patterned flowers.*

ERIA

Eria species

A larger growing member of the Eria *family is* E. sessiflora *with its tall spike of dainty cream flowers that are sweetly scented.*

For more of a challenge try growing erias in a cool to intermediate greenhouse or conservatory, among other similar orchids.

FEATURES

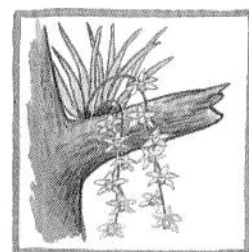

Epiphytic

This is quite a large genus, containing about 500 species, but not many species are grown in cultivation or in amateur collections. They are quite variable in their plant size and habit, but the flowers tend to be of a similar size and shape. *E. sessiflora* is quite a large growing species with a very tall spike of many creamy white flowers; *E. coronaria* in contrast has a short habit with a short swollen stem in place of a pseudobulb but has very similar cream blooms. Many of the erias are scented as well. The family of erias are originally found growing as epiphytes on the trees in the rain forests of the Malaysian Peninsula, the islands of New Guinea and Polynesia as well as some parts of Australia. The plants often have to undergo a wet and dry season in the high altitude monsoon areas so their regime of watering and resting follows this pattern.

ERIA AT A GLANCE

An unusal genus and mostly for the more experienced grower. Charming, small flowers and variable plant size.

JAN	flowering, rest
FEB	flowering, rest
MAR	flowering, rest
APR	flowering, rest
MAY	flowering, water and feed
JUN	water and feed
JULY	water and feed
AUG	water and feed
SEPT	water and feed
OCT	rest
NOV	flowering, rest
DEC	flowering, rest

RECOMMENDED VARIETIES

E. coronaria (cream)
E. javanica (yellow)
E. pubescens (yellow with hairy stems)
E. rosea (pink flush)
E. sessiflora (white)

CONDITIONS

Climate These plants enjoy a cool to intermediate temperature range of 12–28°C (54–83°F) from winter to summer.

Aspect Protect the plants from the bright sun with some shading during summer to prevent the foliage from getting too red or even burnt.

Potting Mix A medium grade of bark will be adequate and you may want to mix in some general potting compost to help keep them moist during the main growing season.

GROWING METHOD

Propagation These orchids are not easily propagated by division as they grow and multiply slowly.

Watering Take care not to over water during the winter when the plant is at rest. Give water once a week during the growing season, keeping the compost moist at all times.

Feeding The plant will respond to added fertilizer during the growing season. Use a soluble fertilizer every two weeks, spraying on the leaves and pouring into the pot.

Problems No specific problems are known if cultural conditions are suitable.

FLOWERING SEASON

Quite short-lived. Blooms for three to four weeks in mainly winter and spring seasons.

GONGORA

Gongora species

The Gongora's *pendant spikes of curious shaped flowers are a constant source of amazement, as seen in this* G. bufonia.

These are easy orchids to grow and are an ideal addition to any cool collection. This G. truncata *is a fragranced example.*

FEATURES

Epiphytic

This fascinating and simple to grow group of orchids originate from the American tropics. They are epiphytic, growing into large clumps over the years. The distinctive ridged pseudobulbs produce pendant flower spikes of varying lengths. The flowers that are held along the length of the thin stem are curiously shaped; the column is elongated and the sepals swept back from it almost like the wings of an insect. This is designed to attract a particular flying insect for pollination. Gongoras are relatively easy to grow in the mixed collection as well as being free flowering and usually scented.

GONGORA AT A GLANCE

Ideal for beginners and growing indoors. Plants are 10–20cm (4–8in) in height and flower spikes reach 40cm (1⅓ft).

JAN	rest
FEB	rest
MAR	water and feed
APR	water and feed
MAY	flowering, water and feed
JUN	flowering, water and feed
JULY	flowering, water and feed
AUG	water and feed
SEPT	water and feed
OCT	rest
NOV	rest
DEC	rest

RECOMMENDED VARIETIES

- *G. bufonia* (cream/red)
- *G. galeata* (orange/brown)
- *G.maculata* (yellow/red)
- *G. quinquinervis* (brown/cream)
- *G. truncata* (pink/cream)

Although flowers only last a few weeks, a mature plant will often produce many flower spikes in succession over the summer months. These orchids are best grown in a basket or net pot.

CONDITIONS

Climate The gongoras are cool growing, needing a drop to 10°C (50°F) in winter. Around 15°C (60°F) in summer is acceptable.

Aspect These orchids have broad, soft green leaves so can easily burn in the sun. Good light in winter and dappled shade in summer.

Potting Mix Needs an open free-draining potting material such as plain bark chippings.

GROWING METHOD

Propagation Once grown into a substantial specimen can be divided up into smaller plants. Only three pseudobulbs are needed for re-flowering.

Watering As the plants grow in open baskets, they will dry out quickly so regular watering is necessary. Immerse the whole plant in water if need be, particularly in summer.

Feeding Only feed when the plant is in active growth, during spring or summer. You can put fertilizer into the water that the plant is being dunked into and leave it for several minutes to let the plant benefit.

Problems No specific problems are known if cultural conditions are suitable.

FLOWERING SEASON

Usually summertime – very free flowering.

HARTWEGIA

Hartwegia species

This is an amazing truly epiphytic orchid. The foliage is beautifully mottled in shades of green and purple.

This Hartwegia purpurea *is always in flower; its wiry stems hold a few bright pink flowers all the year round.*

FEATURES

Epiphytic

This is certainly an interesting orchid for the grower who wants something a little different. The very small genus of *Hartwegia* has only two species from Central America. It is also known as *Nageliella* in some places where the classification has varied. These epiphytic orchids grow well on the smaller branches of tall trees in the lower altitude rain forests of the Central American countries. They are distinctive due to their attractive mottled foliage, thick, leathery leaves and lack of pseudobulbs. The flower spikes are produced from inside a small, green sheath at the base of the leaf and grow very long, up to 1m (3¼ft), before forming buds at the tip. These spikes continue to make more buds throughout the year. With a compact plant and long flower spikes, the *Hartwegia* makes a good orchid to grow on a piece of cork bark in a greenhouse or conservatory.

HARTWEGIA AT A GLANCE

Best suited to growing on bark. Thin flower stems 50cm (20in) long repeat flower all year. Attractive mottled foliage.

JAN	flowering, water and feed
FEB	flowering, water and feed
MAR	flowering, water and feed
APR	flowering, water and feed
MAY	flowering, water and feed
JUN	flowering, water and feed
JULY	flowering, water and feed
AUG	flowering, water and feed
SEPT	flowering, water and feed
OCT	flowering, water and feed
NOV	flowering, water and feed
DEC	flowering, water and feed

RECOMMENDED VARIETIES

H. purpurea

CONDITIONS

Climate These orchids prefer an intermediate climate with a temperature range of 12–25°C (54–60°F) from season to season.

Aspect The tough, thick leaves can be exposed to fairly good light in summer, although, like most other orchids, they will need some protection from the brightest sun. Give as much light as possible in winter.

Potting Mix The habit of the plant ensures that it does not stay inside a pot for very long, so often grows better mounted on bark where it can live undisturbed for many years.

GROWING METHOD

Propagation After some years of growth, a mature plant may be divided into parts, which have at least four leaves, to ensure re-flowering soon.

Watering If growing on a bark slab, the plant will need to be sprayed or soaked in water regularly – every day in summer, less in winter.

Feeding Add feed to the water, and spray this over the whole plant once a week while the plant is producing new leaves. Use a weak dilution of a recommended orchid fertilizer.

Problems If the plant dries out too much on the bark, it will start to dehydrate.

FLOWERING SEASON

All year round, buds come from old stems.

LAELIA

Laelia species

The queen of laelias is Laelia purpurata *and this variety,* carnea, *is especially beautiful with its subtle salmon-coloured lip.*

A cooler growing species is L. gouldiana, *which has long-lasting large pinkish purple flowers on a long, upright stem.*

FEATURES

Epiphytic

There are many different *Laelia* species, coming mostly from Central and the more northern parts of South America. The showy blooms and relative ease of culture make it popular with beginners. It is a varied genus; the size of the plant can vary from 5cm (2in) high, up to 70cm (28in) high, and the flower sizes for these species are similarly different. Colours range from pure white, yellow, lavenders, and pinks to deep purples. They are very closely related to the *Cattleya* family and have been extensively interbred to produce the beautiful *Laeliocattleyas*. *L. purpurata* is perhaps the most well known, being called the queen of laelias. It is the national flower of Brazil and has more cultivated varieties than any other orchid. The laelias have thick, leathery leaves on the top of usually elongated pseudobulbs. Flowers come from inside a sheath at the apex of the newest pseudobulb.

LAELIA AT A GLANCE

Long-lasting, bright flowers, some fragranced. Plants come in a variety of sizes, from 5–45cm (2–18in).

JAN	rest
FEB	rest
MAR	rest
APR	rest, re-pot
MAY	flowering, water and feed
JUN	flowering, water and feed
JULY	flowering, water and feed
AUG	flowering, water and feed
SEPT	water and feed
OCT	rest
NOV	rest
DEC	rest

RECOMMENDED VARIETIES

L. anceps (lavender)
L. autumnalis (lavender)
L. briegeri (yellow)
L. harpophylla (orange)
L. pumila (purple)
L. purpurata (various: white to purple)

CONDITIONS

Climate There are both cool and intermediate growing laelias so check the label. The cooler ones need to drop to 10°C (50°F) in winter, while others need a slightly warmer temperature of 12°C (54°F) minimum.

Aspect Give good light all year round, this is important to encourage flowering. Avoid direct summer sun, which can scorch leaves.

Potting Mix Use a very open, coarse bark mix to make sure the roots are never too wet.

GROWING METHOD

Propagation Although a little slow growing, some laelias will readily propagate after a few years of growing into a larger sized mature plant. Some will shoot from old back bulbs that can be removed at potting time and grown on in a warm, humid place.

Watering Make sure the compost dries out in between waterings. In winter keep dry unless the pseudobulbs start to shrivel.

Feeding Only apply a liquid feed and mist foliage when the plant is actually in growth.

Problems No specific problems are known if cultural conditions are suitable.

FLOWERING SEASON

Most of the laelias are summer flowering, or produce from new pseudobulb in autumn.

LEPANTHOPSIS
Lepanthopsis species

A truly miniature orchid, Lepanthopsis astrophorea *'Stalky', has flowers only a few millimetres across but is seldom out of bloom.*

Although small in stature, a mature specimen will grow into a larger plant measuring 8–10cm (3–4in) across if left undivided for years.

FEATURES

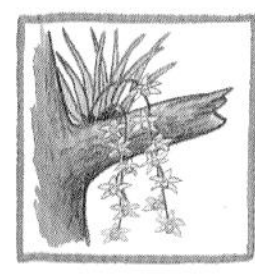
Epiphytic

This is one of several orchid genera, including *Pleurothallis* and *Dryadella*, that is truly miniature. The plant size reaches only 3cm (1¼in) at the most and the flower stem is about the same again. The plant produces a tiny spray of these exquisite little flowers, which measure only 2mm (⅛in) across. The flowers are a vivid deep purple and star-shaped, which makes them a little easier to pick out, however you may still need a magnifying glass. The unusual orchid, *L. astrophorea* 'Stalky', is just one of a genus of around 25 different species widely found in Central America. This genus was originally classified with the *Pleurothallis* until it was proclaimed to be different enough to be given its own name. It grows well with other members of the 'Pleurothallid Alliance' including masdevallias. It can be grown for many years in the same small pot, as it will not outgrow it very easily and will stay almost perpetually in flower.

LEPANTHOPSIS AT A GLANCE

Miniature, only 3cm (1¼in) high but small flowers bloom continually. Best in sheltered, controlled environment.

JAN	water
FEB	water
MAR	water, re-pot
APR	water, re-pot
MAY	water
JUN	water and feed
JULY	water and feed
AUG	water and feed
SEPT	water
OCT	water
NOV	water
DEC	water

RECOMMENDED VARIETIES

L. astrophorea 'Stalky'

CONDITIONS

Climate This orchid is mostly cool growing but will tolerate slightly warmer intermediate conditions if necessary.

Aspect Provide good shade for this little plant so it does not dehydrate too much.

Potting Mix As it is growing in a tiny pot and has a very fine root system, use a fine grade of bark with a little perlite and sphagnum moss.

GROWING METHOD

Propagation Will propagate quite easily once you have let the plant grow on for several years to fill the pot. For best results, though, keep as one plant; it won't take up that much room.

Watering Small pots tend to dry out more quickly than large ones so water regularly to keep from drying out. Mist the foliage also.

Feeding Give the plant a little weak orchid feed during the summer when it is in its more active growth.

Problems No specific problems are known if cultural conditions are suitable.

FLOWERING SEASON

Can be all the year round on a mature plant; generally does not have a strict flowering season. Long lasting for such tiny flowers.

LYCASTE

Lycaste species and hybrids

This orchid, with its delicate pink and white flowers, is derived from Lycaste skinneri, *known to some as the queen of lycastes.*

Large leaves and flowers on single stems are features of the Lycaste *species. This orchid shades prettily from dusky rose to white.*

FEATURES

Epiphytic

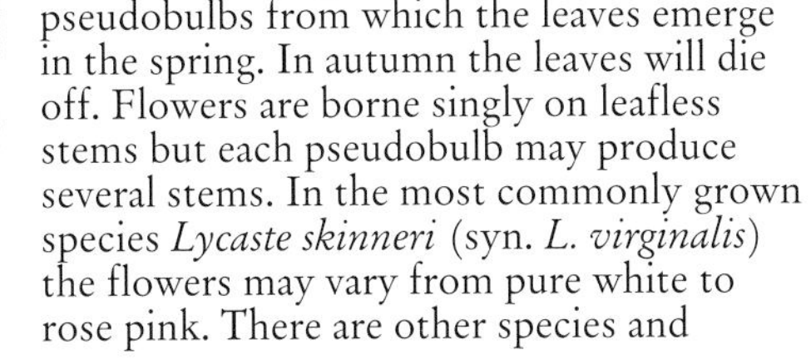

Terrestrial

This group of mostly epiphytic orchids originate in cloud forests in Central America and northern parts of South America. Most are found in the forks of trees but some grow in pockets of leaf litter on rocks. Lycastes can grow into clumps with large leaves and they need space to show to advantage. They produce robust pseudobulbs from which the leaves emerge in the spring. In autumn the leaves will die off. Flowers are borne singly on leafless stems but each pseudobulb may produce several stems. In the most commonly grown species *Lycaste skinneri* (syn. *L. virginalis*) the flowers may vary from pure white to rose pink. There are other species and numerous hybrids with green, yellow or even deep red flowers. Some beautifully shaded or mottled.

For beginners A good species to try is *L. deppei* from Mexico and Guatemala. This is a green flower flecked with red and with a red-spotted yellow lip. It is also fragrant.

LYCASTE AT A GLANCE

Relatively easy to grow but best in greenhouse. Deciduous, with flowers in spring with the new growth.

JAN	rest
FEB	rest
MAR	flowering, water and feed
APR	flowering, water and feed
MAY	flowering, water and feed
JUN	flowering, water and feed
JULY	flowering, water and feed, re-pot
AUG	water and feed
SEPT	water and feed
OCT	rest
NOV	rest
DEC	rest

RECOMMENDED VARIETIES

L. aromatica (yellow)
L. cruenta (yellow)
L. deppei (green/red)
L. skinneri (white/pink)
L. Auburn (pink)
L. Always (pink/red)

CONDITIONS

Climate All are frost sensitive and not ideal for the tropics. Many are happy in a 10–30°C (50–86°F) range. Some species may drop lower during their dry rest.

Aspect Plants require good shade in summer. Early morning sun with shade for the rest of the day is suitable. Ensure good air circulation.

Potting mix Use a mix of crushed fine to medium bark. Do not overpot. Select a pot large enough to take the root ball with a little extra space.

GROWING METHOD

Propagation Divide plants after flowering if they are very crowded. Leafless pseudobulbs can be detached from the clump and potted up.

Watering Give ample water during hot weather and when in active growth but avoid overhead watering. Keep compost dry when dormant.

Feeding Plants respond well to regular fertilizing during the growing season. Use soluble liquids or granular fertilizers. Those formulated for cymbidiums are suitable.

Problems No specific problems are known if cultural conditions are suitable.

FLOWERING SEASON

Mainly winter to early spring.

MASDEVALLIA

Kite orchid

This spotted hybrid Masdevallia *has a pattern of deep magenta markings on a white ground.*

Emerging from a base of heavy leaves, this clear yellow Masdevallia *hybrid has long, very fine tails on its flower.*

FEATURES

Epiphytic

Terrestrial

These unusual flowers are not obviously divided into petals and sepals but appear as solid, sometimes triangular shapes, often with long tails. Flowers may come singly or in small sprays above or within the foliage. The colour range includes white, pink, red, yellow, orange and greenish brown. Some flowers have contrasting venation that looks like stripes. Masdevallias are native to tropical America where they occur from warm lowlands to high altitudes. Plants may be epiphytes, lithophytes or terrestrial and most are found in high elevations in cloud forests. They lack pseudobulbs, growing from a root system that produces short, upright stems, each with a single fleshy leaf.

MASDEVALLIA AT A GLANCE

Miniature orchids, 3–15cm (1¼– 6in) high. Best in cool greenhouse. Flowers 1–3cm (¼–1¼ in) across with long tails.

JAN	rest
FEB	rest
MAR	water and feed,
APR	water and feed, re-pot
MAY	water and feed, re-pot
JUN	water and feed, re-pot
JULY	water and feed
AUG	water and feed
SEPT	water and feed
OCT	water and feed, rest
NOV	rest
DEC	rest

RECOMMENDED VARIETIES

M. barlaeana (red)
M. coccinea (varies)
M. tovarensis (white)
M. Angel Frost (yellow)
M. Marguerite (orange)
M. Whiskers (orange/ purple)

CONDITIONS

Climate This orchid is frost sensitive but classed as cool growing; it prefers a range of about 10–24°C (50–75°F).

Aspect Needs about 70 per cent shade in the summer months; less in winter. Maintain high humidity but keep air moving with fans as good ventilation is needed to grow these little plants well.

Potting mix Use a small pot just large enough for the roots. Use a compost of fine-grade bark mixed with charcoal, perlite and pea gravel.

GROWING METHOD

Propagation This is not easy. Large clumps may be divided after some years. Each division should be made up of at least four stems to make sure they continue to grow well.

Watering Keep the compost moist at all times, especially in warm weather. Growing in small pots, with fine compost, they tend to dry out easily.

Feeding Use soluble liquid plant foods, which can be applied during normal watering in the growing season diluted to half strength.

Problems No specific pest or disease problems are known if the exacting growing conditions are met. If the long flower tails shrivel in warm weather the humidity is too low or the temperature is too high.

FLOWERING SEASON

Flowering depends on species but most flower through spring or summer. When grown to a large plant, masdevallias will provide a lovely show of dainty blooms.

MAXILLARIA

Maxillaria species

Nestling among the foliage are these charming, strongly scented, rich red flowers of the easy-to-grow, cool species Maxillaria tenuifolia.

Another highly fragrant species, M. ochroleuca, *is very free-flowering and easy to grow for the beginner who is just starting to keep orchids.*

FEATURES

Epiphytic

The *Maxillaria* family is a large one, containing around 300 different species, which are found all over the tropical Americas. Their plant and flower sizes vary quite a lot from species to species but the shape of the blooms is very similar, being triangular in shape and likened to the *Lycaste* flowers. Some produce a clump of tightly packed pseudobulbs whereas others, with a more creeping habit, produce a longer rhizome connecting the pseudobulbs. Most species are cool growing and there are many that are relatively easy for the amateur grower. Only a few hybrids have been made between the species. Another bonus is that they are often fragrant, and many produce a cluster of flowers giving a good show. These orchids are always popular and worth having a go at growing in a mixed orchid collection for a cool greenhouse or room.

MAXILLARIA AT A GLANCE

Easy to grow indoors and often scented. Compact plant, 10–25cm (4–10in) high with clustered flowers.

JAN	rest
FEB	flowering, rest
MAR	flowering, water and feed, re-pot
APR	flowering, water and feed, re-pot
MAY	flowering, water and feed
JUN	flowering, water and feed
JULY	flowering, water and feed
AUG	flowering, water and feed
SEPT	flowering, water and feed
OCT	water and feed
NOV	rest
DEC	rest

RECOMMENDED VARIETIES

M. ochroleuca
M. picta (yellow)
M. praestans (brown)
M. rhombea (pale orange)
M. tenuifolia

CONDITIONS

Climate Maxillarias are mostly cool growing and will thrive best in a temperature range of 10–25°C (50–77°F), making sure they receive that cooler drop in the winter. There are a few which would prefer a minimum of 12°C (54°F) at that time of year.

Aspect North facing in the summer to shade the plants from the sun but a south facing or brighter aspect in winter will be preferable.

Potting Mix A medium or fine grade of bark will suit them well; smaller growing species will do better in the finer grade.

GROWING METHOD

Propagation The *Maxillaria* family is very varied so there are some species, such as *M. grandiflora*, which grow slowly. *M. tenuifolia* makes multiple new growths each year so quickly grows into a specimen that can be made into smaller plants if required. For best results leave undivided for as long as possible.

Watering Only water when there are new growths on the plant and then rest, giving only very occasional watering, at other times.

Feeding Only feed when in growth, during summer.

Problems No specific problems are known if cultural conditions are suitable.

FLOWERING SEASON

Varies, but mainly the spring and summer.

Miltoniopsis

Pansy orchid

The Miltoniopsis *are known as pansy orchids due to the shape of these showy blooms, which flower mostly in the summer.*

M. *St. Helier has a contrasting dark mask making up the centre of the flower, which unusually contains no yellow colouring.*

FEATURES

Epiphytic

This is a group of orchids that has gained enormously in popularity in recent years. Commonly known as the pansy orchid, it is easy to see why with their large, showy flowers closely resembling those of the pansies. With extensive breeding the range of colour and patterning of these flowers is becoming endless, and the size and quality getting better and better. With greater availability, this plant is becoming more popular with the amateur and can be grown fairly successfully in the home providing a few particular guidelines are followed. *Miltoniopsis* were once classified with the closely related family of *Miltonia* and it is by this name that they are often still referred to. Now, though, the true miltonias are restricted to just the Brazilian species, which actually have quite a different habit and flower shape to the distinctive Colombian *Miltoniopsis*. The pansy orchids are native to many of the coastal cloud forests of the Colombian mountains and the species grow here on the tree branches alongside other epiphytes such as air plants and bromeliads where the humidity is high.

Some of the original species are still grown and sought after but are not as strong as the modern day hybrids. The plants grow with flattened pseudobulbs from the base of the plant with leaves coming from the sides and top of each one, similar to those of the *Odontoglossum*. The leaves are quite a pale green and soft to the touch so will be easily damaged if handled too roughly. They are in fact closely related to the odontoglossums, which makes them ideal breeding partners, producing some lovely hybrids called odontonias. They have also been bred with other genera such as *Brassia* to give *Miltassia* and *Oncidium* to give a *Miltonicidium*.

MILTONIOPSIS AT A GLANCE

Large showy blooms, up to 10cm (4in) across, make this a popular choice. Often scented, will last for six weeks.

JAN	rest
FEB	rest
MAR	rest
APR	flowering, water and feed
MAY	flowering, water and feed
JUN	flowering, water and feed
JULY	flowering, water and feed
AUG	flowering, water and feed, re-pot
SEPT	rest
OCT	rest
NOV	rest
DEC	rest

RECOMMENDED VARIETIES

- *M. roezlii* (white)
- *M. vexillaria* (pink)
- *M.* Hamburg (red)
- *M.* Robert Strauss (white)
- *M.* St Helier (pink and white)

CONDITIONS

Climate

Miltoniopsis are generally cool growing but they can be susceptible to the cold so it is best to treat them as intermediate growing orchids. This means that they will need a minimum temperature in the winter of 12°C (54°F). Avoid any draughts or chills as these can easily upset the plants. A fluctuation of about 10°C (50°F) from night to day is acceptable so the temperature can rise to 25°C (77°F) in the daytime, especially in summer when it is naturally warmer anyway.

These orchids make good houseplants if given a shaded, humid environment to live in with a little heat in winter.

Pansy orchids, such as this M. Hannover, *are known for their subtle sweet fragrance which is enhanced by warm sunshine and humidity.*

Aspect Miltoniopsis prefer a very shady position to grow in. If there is too much sun then the leaves will become even paler and a yellow tinge will appear. Dappled shade is ideal.

Potting Mix It is best to use quite an open potting mix as the roots of these orchids do not like to be too wet. A medium grade bark is suitable.

GROWING METHOD

Propagation Once the plant has reached a larger size it will regularly produce new pseudobulbs and grow quickly. If the plant is then divided it can be made into plants as small as three pseudobulbs each and they should still flower the next year. However, a large mature plant will look much more impressive as it will have several flower spikes on it at one time.

Watering One thing that Miltoniopsis do not like is to be too wet. They prefer to dry out frequently at their roots between waterings. They also like to be humid so gently mist the foliage every day in warm weather, but a lot less in the cooler winter months, if at all.

Feeding Apply a weak feed to the miltoniopsis when they are making new leaves. This is usually during the spring and summer.

Problems If the compost is constantly kept too wet, this may well cause the problem of root rot, which might then spread up into the pseudobulbs. In extreme cases this can happen very quickly and kill the plant, so to avoid this potential problem make sure that the compost is not kept constantly wet or that the plant is not left standing in water for long periods. When spraying the plant, avoid allowing water to collect inside new growths for long periods of time.

FLOWERING SEASON

The flowering season for these orchids is mainly during the late spring and early summer but plants can bloom out of season if the stage of growth is right. The flower spikes are produced from the base of the completed pseudobulb and the first leaf that covers it. They are long lasting flowers and are quite often fragrant.

Despite the fact that Miltoniopsis flowers are long lasting whilst on the plant, unfortunately they do not live for long when cut. Their soft, papery flowery texture does not last well and will also bruise easily if the plant is moved around a lot. For best results keep this orchid in a shady position, making sure that it is not too hot and the blooms should go on for four to six weeks.

M. Rozel *is one of the deepest purple varieties to be bred recently. The dark, almost black mask contrasts well with its white border.*

2.
3.
7.
8.
1.

6.

5.

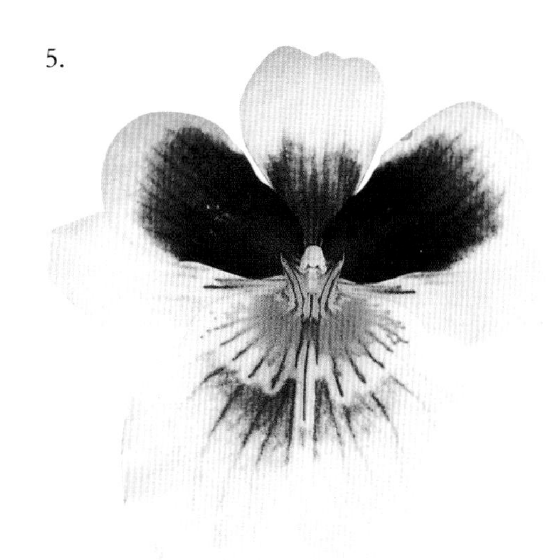

4.

Miltoniopsis

1 Miltoniopsis *Maufant is striking with its contrasting rich pink background and the very deep, almost black mask in the centre of the flower, bordered with white.*

2 M. *Hamburg Dark Tower is a variety well known for its deep red colouring, golden centre and its smooth, velvety texture to the flower.*

3 The two species M. vexillaria *and* M. phalaenopsis *have been crossed together to make this amazing hybrid* M. *Venus, with its stunningly beautiful and unusual waterfall-patterned lip.*

4 M. *St. Mary is a classic cream coloured pansy orchid with only a touch of pink and yellow in the centre, which makes the bloom even more charming.*

5 M. *Jersey is a very large flowered hybrid of this family of orchids, which are becoming more and more popular and easy to grow as houseplants.*

6 A very deep but bright pink is the basis for this lovely pansy orchid, M. *Grin 'Pink Frill', which is reflected in its exceedingly apt name.*

7 This is the species M. vexillaria *from which many of the modern day hybrids have been bred over the years. It still very much keeps its own appeal and popularity because it is a reliable, long-lasting bloomer.*

8 Miltoniopsis *Santa Barbara 'Rainbow Swirl' is a curious variety – when it first opens it is mostly white and over the subsequent days the dark purple blushing increases while the flowers age.*

ODONTOGLOSSUM

Tiger orchid

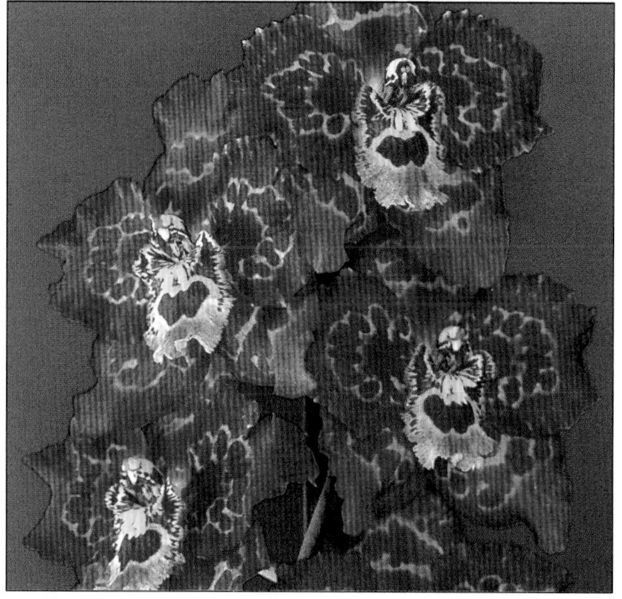

Brilliant colours and intricate patterns are now appearing in many of the newer Odontoglossum *hybrids.*

The darker pattern cleverly mimics the flower shape on this pretty tiger orchid, which makes a good houseplant for a cool room.

FEATURES

Epiphytic

This large group of evergreen orchids comes from Central and South America. They are epiphytes or lithophytes and most occur at high altitudes. They grow from pseudobulbs from which one or two leaves emerge. The flower spikes are very variable and may be short or tall, upright or arching in habit, but all originate from the base of the pseudobulb. Some species are grown but it is the hybrids that are widely cultivated. Hybrids occur in almost every colour of the rainbow and are marked in an extraordinary range of patterns. The odontoglossums will readily breed with other closely related genera to give an ever increasing range of flower types. This interbreeding can also often make the plants much more tolerant to warmer or cooler conditions. Some examples are crossed with *Oncidium* to make *Odontocidium*, with *Miltonia* to create *Odontonia* and with *Cochlioda* to make *Odontioda*. When a third genus is involved then the names change again – for example *Odontoglossum* x *Oncidium* x *Cochlioda* makes a *Wilsonara*.

ODONTOGLOSSUM AT A GLANCE

Popular and easy to grow houseplants. Different varieties with attractive dark green foliage and sprays of flowers.

JAN	less water
FEB	less water
MAR	water and feed
APR	water and feed
MAY	water and feed
JUN	water and feed
JULY	water and feed
AUG	water and feed
SEPT	water and feed
OCT	less water
NOV	less water
DEC	less water

RECOMMENDED VARIETIES

O.crispum (white)
O. cordatum (yellow/brown)
O. hallii (yellow)
O. laeve (brown/white)
Odontocidium Purbeck Gold (yellow/brown)
Odontonia Boussole 'Blanche' (white)
Vuylstekeara Cambria 'Plush' (red/white)

CONDITIONS

Climate Needs a frost-free, but generally not hot, climate and prefers a temperature range from 10–25°C (50–77°F). High humidity is essential for this orchid.

Aspect Needs shade, especially in summer when shadecloth of about 70 per cent should be used. Reduce in winter to provide maximum light to encourage flowering.

Potting mix Needs very open and free-draining soil. A suitable mix would contain medium grade bark, charcoal, pea gravel or very coarse, washed sand. A little chopped sphagnum moss can be added. Do not overpot. Use pots just large enough to confine the roots.

GROWING METHOD

Propagation Divide plants after flowering but only when the container is overflowing. Plants resent frequent disturbance.

Watering Never allow plants to become bone dry but roots should never be sodden. Frequency of watering depends on the mix used and the weather. Mist plants in hot, dry weather.

Feeding Use soluble liquid plant foods at half the recommended strength every couple of weeks through the warmer months.

This beautiful Odontoglossum *hybrid has been bred from the white species* O. crispum *giving it its clear colour and arching spray.*

Bright, golden yellow is a very popular colour in this easy to grow group of orchids which are often chosen as ideal beginner's plants.

Problems No specific pest or disease problems are known for this orchid.

FLOWERING SEASON

The flowering season is very variable, depending on the species or hybrid.

OTHER ODONTOGLOSSUMS TO GROW

O. crispum This species, with pure white flowers, was imported into Britain in large quantities at the beginning of the 20th century, when it commanded enormous prices. Still sought after, it is not found in such abundance as it used to be. Hybrids that have been made with it still retain the beautiful white colouring but tend to be easier to grow, making them good for beginners.

O. cordatum A more compact species which is relatively easy to grow and flower for the amateur enthusiast. Star-shaped golden yellow and chestnut brown flowers are approximately 6cm (2½in) across and between three and six are held on a short arching spray.

O. hallii Larger growing, the plant height being around 30cm (12in) with a spray of large yellow flowers, spotted in brown which have the added bonus of being scented and usually summer flowering.

O. laeve Although this species has small flowers, coloured in dark green and brown with a contrasting white and magenta lip, there are a lot of them held on a very tall, branching flower spike which can be up to 1m (3¼ft) high. The flowers have a strong, sweet fragrance, which will fill your greenhouse. The plant is strong and vigorous growing which makes it good for ease of culture.

There are literally thousands of *Odontoglossum* hybrids available with new ones being produced all the time; here are some examples to choose from.

***Vuylstekeara* Cambria 'Plush'** This is a complex hybrid between three different genera, which make up part of the '*Odontoglossum* Alliance' of related genera. It is one of the all time classic hybrids and has been around since the 1930s and is still very popular today. Its ease of culture and free-flowering habit make it an ideal beginners' orchid as well as being very showy with its large bright red flowers, the lip white with red spotting.

***Odontocidium* Purbeck Gold** This is another classic variety, this time in a brilliant golden yellow, the petals and sepals with just a touch of chocolate brown. Flowers are 8cm (3in) across on a spray reaching 30–50cm (12–20in), depending on the maturity of the plant. These orchids are very easy to grow and flower.

***Beallara* Tahama Glacier 'Green'** The addition of *Brassia* into the breeding of this hybrid gives the flower a stunning, star-shaped appearance. The large blooms can measure up to 6cm (2½in) across. The translucent green of the flower is contrasted with the dark red in its centre. Produces tall sprays with between six and a dozen of these showy, long-lasting flowers. Tolerant of varying temperatures, this orchid will flower well in cool or warm environments. A vigorous grower, it makes an excellent specimen plant in just a few years.

2.
4.
8.
6.
7.

Odontoglossum

1 Sanderara *Rippon Tor 'Burnham' is a very reliable bloomer, always in the late spring and rewards the grower with an attractive arching spray of patterned flowers.*

2 Odontioda *Grenadier is just one of the many red hybrids that include the bright red species* Cochlioda noetzliana *in its family tree giving the striking colour that we see here and in many other red hybrids.*

3 Odontocidium *Purbeck Gold shows the introduction of the genus* Oncidium *into the breeding so giving us the bright yellow influence which also makes the plants very tolerant and easy to grow as houseplants.*

4 A more unusual type is this Colmanara *Wild Cat, which is certainly wild with its leopard spotting and readily branching spikes giving lots of long-lasting flowers. Also cool growing.*

5 Most colours are represented in this widely hybridized family and purple is no exception as seen here in the beautiful Vuylstekeara *Monica 'Burnham'.*

6 The Brassia *influence is clear here with the large star-shaped flowers of the* Beallara *Tahoma Glacier 'Green' which is a robust and free-flowering ideal pot-plant.*

7 Odontioda *Moliere 'Polka' has amazing, frilled edges to the flowers and an extremely intricate patterning in pinks and purples covering the whole flower.*

8 Odontioda *Marie Noel 'Burgogne' is another highly frilled flower which makes a particularly attractive arching flower spike on a mature plant.*

OERSTEDELLA

Oerstedella species

Oerstedellas flower easily on young plants and will happily grow into a clump or in a tray, to give an even better winter show.

The dainty, lilac flowers are long-lasting, continuing for six to eight weeks and will re-flower the following year with relative ease.

FEATURES

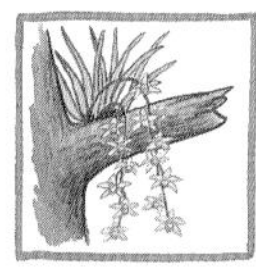

Epiphytic

The genus *Oerstedella* contains species from Central America, from what used to be part of the *Epidendrum* family to which they are closely related and grow in similar conditions. *O.centradenia* is popular for its ease of culture and is very free flowering. The thin, reed like growth does not get quite as tall as some of the larger growing epidendrums and has the habit of producing lots of keikis. These are plantlets, which grow on the side of the main stem and will flower well at a very small size. These, as well as the parent plant, make a lot of aerial roots so they will grow well either in a pot or on a piece of cork bark. The clusters of small, pretty pink flowers are produced from the top of the last season's growth and are extremely long lasting, remaining perfect on the plant for up to six weeks.

OERSTEDELLA AT A GLANCE

Compact and easy to grow with long lasting dainty flowers on a regular annual basis. Easily propagated.

JAN	flowering, rest
FEB	flowering, rest
MAR	flowering, rest, repot
APR	water and feed, re-pot
MAY	water and feed
JUN	water and feed
JULY	water and feed
AUG	water and feed
SEPT	water and feed
OCT	rest
NOV	rest
DEC	rest

RECOMMENDED VARIETIES

O. centradenia (pink)

O. centropetala (green/ purple)

O. endresii (white/ purple)

CONDITIONS

Climate This orchid does best in the intermediate to warm section of the orchid house, needing a range of 15–28°C (60–83°F) to thrive.

Aspect Shade from the brightest summer sun to avoid the small leaves from becoming too pale. Give maximum light in the winter.

Potting Mix Use an open mixture to accommodate their thick roots, a combination of fine grade bark and perlite would suit them very well.

GROWING METHOD

Propagation The easiest way to propagate this orchid is by removing the keikis that are made automatically from the sides of the main stems and potting them up separately. Do this when they have developed their own root system and are several centimetres high. The resulting plants will flower at a very small size, probably the following year after removal from the parent plant.

Watering Water and spray regularly throughout the year but especially in the summer.

Feeding Add feed to the water that you give during the summer, every other week.

Problems Make sure you mist the leaves regularly as they can be prone to red spider mite if they are kept too dry on the surface.

FLOWERING SEASON

These are winter flowering, and will flower regularly every year.

ONCIDIUM

Dancing ladies

The scarlet flaring lip of this Oncidium *hybrid, Hawaiian Gold, resembles a flamenco dancer's skirt.*

Native to Brazil, the species Oncidium sarcodes *has each of its flowers neatly spaced along the stem.*

FEATURES

Epiphytic

This group of orchids originates in tropical and temperate areas of central and South America. Most in cultivation are evergreen and epiphytic. Yellow flowers with dark to red-brown markings are most common but *Oncidium ornithorhynchum* has mainly pink, scented flowers. Flowers in the species vary from tiny gems 1cm (½in) wide to 8–10cm (4–5in). Flowers are produced in sprays on the ends of stems. The species *O. flexuosum* was originally given the name of 'dancing doll' to describe the dainty yellow flowers on the ends of fine stems which bob about in a breeze. Some have a flattened pseudobulb with one or two leaves; others are compact with a fan of foliage.

ONCIDIUM AT A GLANCE

Easy to grow so perfect for beginners, with showers of brightly coloured flowers on long sprays up to 70cm (28in).

JAN	rest
FEB	re-pot, rest
MAR	re-pot
APR	water and feed
MAY	water and feed
JUN	water and feed
JULY	flowering, water and feed
AUG	flowering, water and feed
SEPT	flowering, water and feed, rest
OCT	flowering, water and feed, rest
NOV	flowers, rest
DEC	flowers, rest

RECOMMENDED VARIETIES

O. flexuosum (yellow)
O. ornithorhynchum (pink)
O. sphacelatum (yellow)
O. Gower Ramsey (yellow)
O. Star Wars (yellow)

Habit Some grow upwards. In pots provide a pole or slab, and stones for stability for those that need to be trained. They can also be grown in hanging baskets or tied to slabs.

CONDITIONS

Climate Depends on the origin of the plant. Some come from cool mountain areas with low rainfall but almost permanent cloud or mist, others from hot, steamy lowland areas. Some prefer distinct wet and dry seasons. Seek advice before buying.

Aspect Needs good light but not direct sun.

Potting mix Perfect drainage is essential. In containers, use coarse bark for potting. Cork slabs make good bases.

GROWING METHOD

Propagation Divide plants overflowing their containers or remove backbulbs when present and pot them up separately after flowering.

Watering Watering depends on the season but in general water frequently in summer and allow to dry out between waterings in winter. Maintain humidity. In cool weather plants may often need only misting.

Feeding Use liquid fertilizers only during periods of active growth at half recommended strength.

Problems Fungal leaf spots can be a problem. If so, increase ventilation and avoid watering foliage. Water early in the day.

FLOWERING SEASON

There is a species in flower most months of the year but many flower in autumn to spring. Large plants can flower throughout the year.

PAPHIOPEDILUM

Slipper orchid

The shiny pouches on these slipper orchids look wax-like and almost artificial. The flower structure is most intricate.

Rising above a cluster of mottled leaves are the purple striped and spotted flowers of this well-shaped slipper orchid.

FEATURES

Terrestrial

Slipper orchid flowers are very beautiful to some people while others can see no attraction in them. They are certainly fascinating in their enormous variety of shapes, sizes, colours and patterns but many of the newer hybrids, especially, look almost artificial. The species tend to be more finely formed than many of the modern large flowers with a rounded outline. The long-lasting flowers are usually solitary among the dark green or sometimes mottled leaves. Flower colours include green, white, yellow, red, purple, pink and brown, mostly with at least two colours on each bloom. Many are striped or spotted and all have the distinctive pouch-shaped lip, giving rise to the common name. These orchids are mostly terrestrial although a few are lithophytes, growing on rocks.

Habitat The species are mainly native to south-east Asia, with some occurring in the Himalayas and areas through China and extending south to New Guinea. The range of habitats is diverse, too, and different species are found from sea level to mountainous areas with very high altitudes.

Availability Many species of *Paphiopedilum* are rare and threatened in the wild and so few are offered for sale. There is, however, an almost endless variety of hybrids available from specialist growers and collectors. It is worth visiting specialist nurseries when plants are in bloom or attending orchid shows to get an idea of the huge choice within this orchid group before you buy.

Habit Slippers have no pseudobulb and grow as clumps of leafy shoots. When large enough these can be divided. This can make them expensive to buy as they cannot be mass cultured readily and are slow to produce.

Species Species tolerating the most cold include *P. insigne*, *P. venustum* and *P. villosum*. These are often available for sale and they are quite easy to cultivate.

PAPHIOPEDILUM AT A GLANCE

Some types are easy for beginners, others need specialist conditions. Range from 20–50cm (8–20in) high.

JAN	water and feed
FEB	water and feed
MAR	water and feed
APR	water and feed
MAY	water and feed
JUN	water and feed
JULY	water and feed
AUG	water and feed
SEPT	water and feed
OCT	water and feed
NOV	water and feed
DEC	water and feed

RECOMMENDED VARIETIES

- *P. insigne* (copper/green)
- *P. primulinum* (yellow)
- *P. spicerianum* (white/green/purple)
- *P.* Jac Flash (purple)
- *P.* Jersey Freckles (red/white)
- *P.* Maudiae (green/white)
- *P.* Pinocchio (yellow)
- *P.* Roth-Maud (purple/brown)

CONDITIONS

Climate Needs a frost-free climate. Those with plain green leaves are cool growers preferring a temperature range of 10–25°C (50–77°F). Varieties with mottled leaves prefer warmer

The creamy yellow flowers found on the species Paphiopedilum concolor *are patterned with fine purple spots.*

The dark colouring of this orchid gives it an almost sinister look but the striped sepal and twisted petals are very graceful.

growing conditions and are best between 20–30°C (68–86°F). Both of these types like fairly high humidity.

Aspect Needs filtered light, but no direct sun except early in the morning, and more shade than cymbidiums. Must be protected from the hottest summer sun. In winter they need as much bright light as possible.

Potting mix A proprietary orchid mix suitable for terrestrials is usually adequate or this orchid can be grown in small grade pine bark with added limestone chips. The mix must be free draining or roots will rot. If limestone chips are unavailable, soak the bark in limewater for a few hours before using it: 10g (¼oz) of lime or dolomite per litre of water makes a good solution.

Grow this orchid in a pot that is just large enough to confine the roots. You will need to re-pot the plant every couple of years as breakdown of the potting mix often impedes its free-draining character.

GROWING METHOD

Propagation Offshoots of the parent plant may be separated once they have developed good root systems. Avoid dividing until the plant has formed a clump as it may not flower until there are multiple growths. Division is best carried out straight after flowering.

Watering In spring and summer water freely but allow to become almost dry between waterings in winter. It should never dry out completely. In the cooler weather water early in the day and avoid overhead watering so that water is not held in the folds of the leaves, as this encourages mould growth which can lead to the total loss of flower buds or new leaves.

Feeding Feed in the growing season with soluble liquid fertilizer. Plants that are grown where the temperature is controlled can be fed occasionally during the cooler months if they have not gone dormant.

Problems Slugs and snails may cause damage to foliage or flowers. Dead foliage should be cleaned out regularly to avoid harbouring pests. Overwatering or heavy potting mixes cause root rot and possibly the death of the plant. Glasshouse-grown plants can be susceptible to red-spider mites and mealybugs.

Support Stems tend to droop slightly so many growers use wire to hold stems upright as buds develop and stems elongate.

FLOWERING SEASON

Most flower through late autumn, winter and spring. Once the flowers have faded, cut off the stems as low as possible on the plant.

Indoors A potted plant in bloom may be brought into the house where it will look decorative for several weeks.

Cut flowers The slipper orchid may be used as a cut flower, although the display will last far longer if left on the plant.

PHALAENOPSIS
Moth orchid

Red speckling on a lime green background makes for very unusual colouring on this moth orchid.

The name given to this Phalaenopsis *hybrid, Dendi Flares, describes the colouring, with rose-purple flushes over white.*

FEATURES

Epiphytic

For sheer breathtaking beauty there is probably no other orchid to rival the moth orchid. The flowers last for many weeks on the plant. There are about ninety species of moth orchid native to the tropical regions of south-east Asia, the Himalayas, through the Philippines to New Guinea and Australia. Some enthusiasts grow the species but more people grow some of the thousands of hybrids that have been developed by breeders throughout the world. Moth orchids are epiphytic, growing mainly on trees in their native habitats.

PHALAENOPSIS AT A GLANCE

Ideal beginner's orchid, flowering all year round. Flowers vary in size, stems 15–50cm (6–20in)

JAN	water and feed
FEB	water and feed
MAR	water and feed
APR	water and feed
MAY	water and feed
JUN	water and feed
JULY	water and feed
AUG	water and feed
SEPT	water and feed
OCT	water and feed
NOV	water and feed
DEC	water and feed

RECOMMENDED VARIETIES

- *P. amabilis* (white)
- *P. equestris* (pink)
- *P. schilleriana* (pink)
- *P.* Chamonix (white)
- *P.* Cote d'Azur (white/ pink)
- P. Mystik Golden Leopard (yellow)
- *P.* Pink Twilight (pink)
- *P.* Yellow Treasure (yellow)

Foliage The leaves of the moth orchid are slightly fleshy and they may be medium to dark green or even sometimes a mottled colour. Individual *Phalaenopsis* plants may have anywhere from two to six leaves.

Flowers The flower stems may be upright or arching and pendulous and they carry anywhere from one or two to fifty or sixty blooms. Stems may be from 30cm (12in) to over 1m (3¼ft) high. The range of flower colour in the species includes white and almost every other colour except blue. Some of the flowers are striped, barred, spotted or mottled. The cultivated hybrids also contain a huge colour range including those with stripes, bars and blotches, but the most popular hybrids in the recent past seem to be the pure whites with a contrasting lip and those in all the shades of pink from palest pastel to deep rose. Breeders of moth orchids, however, are now producing many with spots, stripes or contrasting shadings, including new yellow varieties.

Cutting Cut moth orchid flowers are very popular for bridal bouquets, buttonholes and table decoration, often as single stems holding several blooms, and will last well in water.

P. amabilis *P. amabilis* is the best known species and is a parent of many modern hybrids. It has plain green leaves and bears numerous white flowers. Flowers appear from autumn to spring. Although this orchid was first discovered in 1750 there were no hybrids registered until 1920. However, its hybrids now number in the thousands. The natural range of this species extends from the Philippines through to north-east Australia.

Finely drawn stripes in an intricate pattern give this moth orchid beautiful detail that will repay close observation.

This lovely truss of white moth orchids with a deep red lip will give many weeks of pleasure as the flowers are long lasting.

P. cornu-cervi *P. cornu-cervi* is distinctive with star-shaped waxy flowers which may be yellow or yellow-green, heavily spotted or blotched with red-brown. The leaves are plain green and the flowering stem may grow to around 40cm (16in). This species is originally from Burma and south-east Asia.

P. schilleriana *P. schilleriana*, which is also from the Philippines, has broad, fleshy, very long leaves that are dark green with silver spots above and purple beneath. The rosy pink flowers are carried on stems up to 1m (3¼ft) long and appear in winter and spring. The combination of decorative foliage and very lovely flowers makes this a popular species.

CONDITIONS

Climate Grows best in warm conditions. The preferred temperature range is from 16 to 30°C, (62–86°F) although temperatures can be much higher as long as a high level of humidity is maintained.

Aspect Needs bright filtered sunlight with highest light levels through autumn and winter. This orchid grows especially well as a house plant, coping particularly well with the modern home and the warmer, insulated environment that comes with it.

Potting mix Grow plants in pots just large enough to contain roots. A mix of coarse bark and charcoal would be suitable. Plastic pots are most often used but a plant with a heavy head of numerous flowers may need to be weighted with pebbles. Large plants of show-size can be grown in terracotta pots to prevent them tipping over.

GROWING METHOD

Propagation Propagation can be slow. Offshoots may develop and they can be separated from the main plant during the warmer months once roots have grown. Most moth orchids are propagated by tissue culture or from seed, both specialist laboratory techniques.

Watering These plants need watering frequently in summer and perhaps less often in winter. Never let the roots become bone dry or sodden. Lightly mist the plants in summer, although water must not be allowed to settle for long periods in the central crown.

Feeding Apply soluble liquid plant foods at half the recommended strength weekly in summer, every two or three weeks in the cooler months. If warm temperatures and high humidity are maintained these orchids do not go through a dormant or resting phase.

Problems No specific pest or disease problems are known. Plants grown in glasshouses may be susceptible to mealybugs and mites. Slugs and snails are always a problem in the damp conditions needed for good growth.

FLOWERING SEASON

Many moth orchids can flower almost all the year round. Some of the straight species have definite flowering seasons. After most of the flowers have fallen from the spike this can be cut down to within the second node from the base. This will often encourage a secondary spike to emerge.

Cut flowers If cutting the original spike for decoration, cut close to the base of the plant.

1.
7.
6.
2.
5.

3.

4.

8.

Phalaenopsis

1 Phalaenopsis *Lady Sakara is as elegant as the name suggests with clear candy striping in deep pink on a snow-white background contrasting with the dark lip.*

2 P. *Alice Girl has a lovely contrast between bright white petals and sepals and the deep crimson lip in the centre.*

3 P. *Pink Twilight is another one of the most popular moth orchids to grow and is an ideal plant for indoors, enjoying the warm, shady conditions of the home.*

4 Many of the phalaenopsis have large flowers on tall sprays but this one, P. MystikGolden Leopard, has numerous smaller blooms on a compact, branching stem, making it a charming addition to any collection.

5 Yellow colouring in this family is not as common as the pinks and whites so this P. *Yellow Treasure is very special and sought after. It is grown in the same conditions as all other types of* Phalaenopsis.

6 P. *Heverlee has very subtle deep pink veining overlaid on a rich purple pink, which makes it a particularly showy orchid to grow.*

7 Like all of the moth orchids, this P. *Follett will reflower from the same stem by cutting it back to just above the highest node on the flower spike after the first flowers have faded to encourage re-branching.*

8 P. *Lippegruss has large flowers typical of most modern* Phalaenopsis *hybrids reaching around 8cm (3in) across and lasting for months in perfect bloom.*

PHRAGMIPEDIUM

Slipper orchid

The colouring of this Phragmipedium *Eric Young is inherited from the red of the parent* P. besseae, *a relatively newly discovered species.*

The compact habit of this species, P. pearcei, *makes it ideal for the grower without much space left in the warm section of the greenhouse.*

FEATURES

Terrestrial

A really interesting and intriguing group of orchids, the phragmipediums are a close relative of the *Paphiopedilum* which used to be classified all together with the *Cypripedium* as one large family of slipper orchids. They do have quite a different appearance but also have a lot in common with their cousins such as their plant structure which consists of broad leaves growing from a central point and side shoots which increase the size of the plant over the years. The flowers come from the centre of the latest set of leaves and they are produced in succession, the next bud opening as the previous flower drops. Some species such *P. longifolium* can grow flower spikes up to 2m (6½ft) high and will be continually in flower for over 18 months. The species originate from all over South America growing in the ground in moist or even boggy areas near rivers and streams where their roots hardly ever dry out. With increased hybridizing over recent years to produce some spectacular colours the plants are also becoming more tolerant and are proving to be relatively easy to grow for the amateur under warm conditions.

PHRAGMIPEDIUM AT A GLANCE

Warm growing plant that can bloom at any time of year. Can reach 50cm (1¾ft) with 4–10cm (1½–4in) flowers.

JAN	water and feed
FEB	water and feed
MAR	water and feed
APR	water and feed, repot
MAY	water and feed, repot
JUN	water and feed
JULY	water and feed
AUG	water and feed
SEPT	water and feed
OCT	water and feed
NOV	water and feed
DEC	water and feed

RECOMMENDED VARIETIES

- *P. besseae* (red)
- *P. longifolium* (green/brown)
- *P. pearcei* (pale green)
- *P. schlimii* (pink)
- *P.* Don Wimber (orange/red)
- *P.* Eric Young (orange)
- *P.* Living Fire (red)
- *P.* Sedenii

CONDITIONS

Climate The phragmipediums are warm growing, like the paphiopedilums, needing a minimum of 15–18°C (60–66°F) in winter and can take up to 28–30°C (83–86°F) in summer. They are quite cold sensitive so it is important to keep the temperature up in winter.

Aspect Good shade is needed as these terrestrial plants grow naturally under a lot of tree cover in the rain forests. The leaves can be scorched by bright sun but not in winter when the rays are not as strong.

Potting Mix Use a slightly heavier mix than is usually used for the epiphytic orchids to keep the roots moist all the time. Make up a mixture of fine bark, perlite and peat or a similar substitute. This will help to keep the moisture around the roots.

GROWING METHOD

Propagation Phragmipediums do not divide easily, as they need to grow fairly large before distinct pieces can be separated. Make sure there are at least three sets of leaves to each part when splitting to ensure good growth.

Many colours are represented in this genus and pink is no exception as seen in this P. Sedenii*, which will bloom for many months.*

One of the most spectacular hybrids is P. Grande *with its extremely long petals hanging down either side of the slipper-shaped pouch.*

Watering Keep these orchids moist at all times, sometimes the plant may need watering every day if it is very warm weather and it is using up the moisture quickly. At that time of year or if the plant is going to be left for some time then stand it in a saucer of water to prevent drying out. If misting the foliage, avoid water collecting in the centre of the new shoots as this can cause rotting.

Feeding Add some weak fertilizer to the water once a week in the main growing season when new leaves are being produced. This is usually more in the summer but as they are warmth loving, they can actually grow at any time of year.

Problems Ensure that the plant is kept moist enough to avoid dehydration but not so wet that the roots actually die and rot.

FLOWERING SEASON

This can be at any time of the year and although the individual flowers only last a short time they will continually re-flower over many weeks.

OTHER PHRAGMIPEDIUMS TO GROW

P. longifolium This is one of the largest growing species, having very long leaves and tall flower spikes holding a succession of green and brown slipper shaped flowers which can go on producing flowers for over a year.

P. besseae A relatively newly discovered species from Peru which, with its bright red flowers, has sparked a whole new breeding line producing stunning hybrids in reds and oranges which have proved to be very popular. Flowers are produced in succession over many weeks.

***P.* Eric Young** This is one of those hybrids, a cross between the two species mentioned above, and inherits the lovely orange colouring but in a larger sized flower, 8cm (3in) across. It is proving to be slightly easier to grow than the species so is increasing popular. Flowers are short-lived but more are produced over a period over several months.

***P.* Sedenii** This is a very early hybrid from the species *P. schlimii*, producing unusually coloured flowers with a pink slipper-shaped pouch.

P. pearcei This miniature growing species has shorter, thin leaves and a creeping rhizome between shoots. The flower spikes reach 10cm (4in), with a succession of green, striped flowers.

P. longifolium *has very elegant flowers with lovely red shading on the long petals and the pouch.*

PLEIONE

Pleione species and hybrids

The small treasures that make up the Pleione *species are often best grown as container plants where they can be enjoyed at close range. Among the features to watch out for is the contrasting lip of the flower, which is delicately fringed.*

FEATURES

Terrestrial

Epiphytic

Extremely cold tolerant, these orchids are native to cool areas of northern India, and to southern China and Taiwan where they are usually found in damp woodland. They are deciduous and may be terrestrial, epiphytic or lithophytic. They die down completely in winter and renew the small pseudobulbs in spring. Plants are rarely more than 15cm (6in) high and may spread to about 30cm (12in). They grow in very shallow pots. Flowers appear before the leaves and are most often white, pink, mauve, purple or even yellow. The lip may be fringed and spotted. Individual blooms are not long lasting but make a fine display when planted en masse. Each small pseudobulb flowers only once and then produces a folded, elliptical leaf. Pleiones make very easy orchids for beginners and especially for enthusiast children. Their compact habit and ease of culture make them ideal house plants for a very cool windowsill or greenhouse. Some alpine growers even include them in their collections. The pots can be placed out of doors for the summer months.

PLEIONE AT A GLANCE

Very easy to grow and ideal for greenhouse. Can reach 15cm (6in) high and flowers average 5cm (2in) across.

JAN	rest
FEB	repot, flowering
MAR	flowering, water and feed
APR	flowering, water and feed
MAY	water and feed
JUN	water and feed
JULY	water and feed
AUG	water and feed
SEPT	flowering, rest
OCT	flowering, rest
NOV	rest
DEC	rest

RECOMMENDED VARIETIES

P. formosana (lilac pink)
P. formosana var. *alba* (white)
P. maculata (white/red)
P. praecox (pink)
P. speciosa (cerise)
P. Eiger (lavender)
P. Piton (lilac)
P. Shantung (yellow)
P. Stromboli (pink)
P. Versailles (pink)

CONDITIONS

Climate Some are frost hardy, others frost tender but all prefer cool to cold conditions. A minimum temperature in winter of 5°C (10°F) is acceptable to these plants as long as they are in their leafless, dormant phase.

Aspect Needs a sheltered spot in filtered sunlight with shade during the hottest part of the day. Glasshouse-grown plants will need heavy shading and cooling in summer.

Potting mix The mix of bark, pea gravel, small pebbles and chopped sphagnum moss or fibre peat must be perfectly drained.

GROWING METHOD

Propagation Plants need re-potting annually at the end of winter when old pseudobulbs can be discarded. It is then, in the early spring, that the new shoots are just starting to grow, with the flower buds inside. As well as removing last year's dead pseudobulb, trim back the dead roots when re-potting. Some varieties are likely to break double each year so the number of pseudobulbs multiply easily over the years. Occasionally, small extra pseudobulbs can be formed at the top of the old, shrivelled bulb. These can be

There are many hybrid pleiones which add to the range of shades of pink, white and yellow including *this striking* Pleione *Soufriere.*

Hybrid Pleione *Shantung is just one of the easy growing orchids that are ideal beginner's plants for the cool greenhouse or conservatory.*

removed at potting time and planted in with the larger main bulb. In a few years time, these will be large enough to flower.

Watering Keep soil moist once growth has started in spring. Water regularly during flowering and the development of the pseudobulbs. These orchids make fine root systems which can dry out easily if not regularly watered. When pseudobulbs are fully matured, reduce the frequency of watering. Keep plants dry in winter while dormant and leafless. To avoid resting pseudobulbs from becoming too wet during winter remove from their pots after their leaves have fallen and leave to dry out in an empty tray or pot. You can also see clearly when the new growth starts and can then pot them up again.

Feeding Give regular, weak liquid fertilizer once growth commences and continue until pseudobulbs are well matured.

Problems Slugs and snails are a constant problem but there are no other specific pest problems. Plants can die from root rot if they are constantly wet while dormant or if they dry out completely during the growing season when they can dehydrate.

FLOWERING SEASON

Most species flower in spring but some flower in autumn.

OTHER PLEIONES TO GROW

P. formosana Probably the most popular *Pleione* species to grow, the number of bulbs multiplies up quickly over the years so a superb show can be achieved in quite a short time. Soft lavender pink petals and sepals with dark pink and brown spotting on the white lip.

P. formosana* var. *alba A pure white, albino form of the above species which has only a touch of yellow in the centre of the lip. The pseudobulbs are also devoid of any purple colouring and are a clear apple green. Grows slightly smaller than the pink variety.

P. speciosa Known for its very vibrant cerise coloured flowers which will brighten up the early spring months when it is in flower.

P. maculata An unusual species as this one flowers in the autumn, one of only two species that does. White blooms, with dark red patterning in the lip, are also unusual in a mostly pink dominated genus.

P. praecox The second autumn flowering species, this time in traditional pink. These two grow in just the same conditions as the spring flowering types but are perhaps a little more of a challenge.

A few hybrid pleiones have been bred between the species; the following are a few examples which are easy to grow:

***P.* Eiger** One of the first to flower in the spring season, short stem with a pale lavender flower, very pretty and easy to keep.

***P.* Piton** A very large sized flower in comparison to the others, 6cm (2½in) across, on a taller stem, 10cm (4in) high. A lovely subtle purple shade with bold spotting on the lip.

***P.* Shantung** One of the most well known of the hybrid pleiones due to it being a yellow hybrid, the darkest form being *P.* Shantung 'Ducat'. Grows well but may not multiply as quickly as some of the others. The most commonly seen variety is *P.* Shantung 'Ridgeway' AM/RHS, a soft yellow with a pink blush.

ROSSIOGLOSSUM

Clown orchid

This bright yellow and red flower is nicknamed the clown orchid after the distinctive character of the bloom.

The flowers of the Rossioglossum *are showy and lasting, especially this easy-to-grow hybrid* R. *Rawdon Jester.*

FEATURES

Epiphytic

This incredibly showy and fascinating orchid was originally part of the *Odontoglossum* family. Probably the best known of the species is *R. grande*, commonly known as the clown orchid. A little man in a colourful yellow and red outfit can be seen at the top of the lip. A few hybrids have been made between some of the species such as R. Jakob Jenny (*grande* x *insleayi*) and R. Rawdon Jester (*grande* x *williamsianum*). The pseudobulbs are oval-shaped and a handsome green with a pair of large, broad leaves in the same colour. The flowers tend to be very long lasting with quite a waxy texture and reaching an amazing 15cm (6in) across from petal tip to petal tip.

ROSSIOGLOSSUM AT A GLANCE

Good for beginners in the home or conservatory. Flowers reach 15cm (6in) across and spike 30cm (12in) above foliage.

JAN	rest
FEB	rest
MAR	rest
APR	rest
MAY	flowering, water and feed
JUN	flowering, water and feed
JULY	water and feed
AUG	water and feed
SEPT	flowering, water and feed
OCT	flowering, rest
NOV	rest
DEC	rest

RECOMMENDED VARIETIES

R. grande
R. insleayi
R. williamsianum
R. Jakob Jenny
R. Rawdon Jester
(all yellow/brown)

CONDITIONS

Climate Traditionally a popular orchid to grow in a cool greenhouse or conservatory, with a minimum temperature of 10°C (50°F) in winter and 10°C (50°F) in summer.

Aspect The leaves prefer a shady position so protect from bright sun. Dappled shade is preferred; a north facing aspect in summer is ideal and south facing in winter.

Potting Mix A fairly open mix is ideal, bark based with some peat or similar mixed in.

GROWING METHOD

Propagation Rossioglossums are quite slow to grow and take many years to reach an easily dividable plant. Therefore, leave the plant until the pot is full before moving to a larger size and only split when necessary and possible.

Watering Likes a well-defined resting period in winter, which goes on well into spring. Water regularly from the point when the new growth starts increasing over the growing season and decrease to a stop in the autumn. Allow compost to dry out a little in winter.

Feeding Use a half strength general plant food every two or three waterings in the growing season.

Problems If a dry rest period is not observed then the plant can suffer from over-watering in the winter which can lead to root rot. Avoid spraying the foliage in winter as this can lead to spotting which can cause fungal infection.

FLOWERING SEASON

R. grande traditionally flowers in autumn but some hybrids, such as *R.* Rawdon Jester will bloom easily in late spring and summer.

SARCHOCHILUS

Orange blossom orchid

This orchid, the dainty little 'Pink Blossom', is a cross between Sarchochilus hartmannii *and* S. falcatus.

Deep red blotches on small white crystalline flowers are a feature of Sarchochilus falcatus *'Pinky'.*

FEATURES

Epiphytic

These small epiphytic or lithophytic orchids are native to Australia with one from New Caledonia. They rarely exceed 15cm (6in) in height and the flower spikes are slightly pendulous, holding three or more flowers. They are generally cool growers but their natural range is more intermediate. The mid-green leaves are lance-shaped and slightly fleshy, and the flowers emerge from the base of the leaf. They are pure white, but may have red bands or dots in the central part. The lip may be marked yellow, orange or red. Some have a very light perfume. *Sarchochilus falcatus*, *S. hartmannii* and *S. fitzgeraldii* are the most common, with some hybrids.

SAROCHILUS AT A GLANCE

Compact plant standing only 10–15cm (4–6in) high with sprays of small 2cm (¾in) flowers.

JAN	rest, water less
FEB	flowering, water and feed
MAR	flowering, water and feed
APR	flowering, water and feed
MAY	flowering, water and feed
JUN	water and feed
JULY	water and feed
AUG	water and feed
SEPT	water and feed
OCT	rest, water less
NOV	rest, water less
DEC	rest, water less

RECOMMENDED VARIETIES

S. falcatus
S. hartmannii
S fitzgeraldii
S. Fitzhart

CONDITIONS

Climate Needs a temperature range from about 12–25°C (54–80°F).

Aspect Prefers some shading during the summer but overall likes good light all year round.

Potting mix A potting mix suitable for cymbidiums can be used as a base. Add charcoal, coarse bark and perlite to keep the mix very open.

GROWING METHOD

Propagation Grow fresh plants from divisions removed from the parent plant in spring or early summer. Do not divide plants too often – wait for them to fill their pots.

Watering Keep roots just a little moist at all times. Give plenty of water during spring and summer but water less frequently through the cooler months. Some of these orchids grow naturally in the spray of waterfalls or on permanently dripping rocks, making them ideal for growing near a water feature. They produce lots of aerial roots which can be sprayed to keep moist. These roots are not always a sign that the plant needs re-potting.

Feeding Use small quantities or weak solutions of fertilizer regularly through the growing season. Slow release granules or half strength liquids are suitable.

Problems No specific problems are known if the cultural conditions are suitable.

FLOWERING SEASON

Produces masses of white flowers with an almost crystalline appearance sometime during spring. Cut off the flower stem once the blooms have fallen.

SOPHRONITIS

Sophronitis species

Deep scarlet flowers are characteristic of Sophronitis coccinea *and its cultivars. Best cultivated in small shallow pots, these orchids look best when plants have multiplied into sizeable colonies. Do not be tempted to divide them too often.*

FEATURES

Epiphytic

These small epiphytic orchids are native to Paraguay and eastern Brazil where they grow in humid cloud forests. They rarely grow more than 5cm (2in) high and spread to make small mats about 10cm (4in) wide. Their oval pseudobulbs are often found in clusters, each bearing a solitary small, leathery, dark green leaf. Flowers are brilliantly coloured and mainly scarlet, orange or yellow although there are species with rosy pink to violet blooms. Their rich colour is a component of many of the hybrids raised with related groups. Plants are generally easy to manage and may be grown in small pots or as epiphytes. The species most often grown is *Sophronitis coccinea*. *S. cernua* has more orange-scarlet flowers and is harder to find.

S. coccinea *S. coccinea* has bright scarlet to orange or yellow flowers. The named cultivars generally have very bright red flowers and seem to be more numerous than those of other species.

SOPHRONITIS AT A GLANCE

Grows best in a greenhouse or conservatory. Compact species, 5cm (2in) high with brilliant red flowers.

JAN	rest
FEB	rest
MAR	flowering, rest
APR	flowering, water and feed, re-pot
MAY	flowering, water and feed, re-pot
JUN	water and feed, re-pot
JULY	water and feed
AUG	water and feed
SEPT	water and feed
OCT	rest
NOV	rest
DEC	rest

RECOMMENDED VARIETIES

S. cernua (red-orange)

S. coccinea (bright red)

CONDITIONS

Climate Needs a frost-free but not tropical climate. Most prefer a temperature range of about 10–30°C (50–86°F) and high humidity.

Aspect These orchids need light shade and good air circulation.

Potting mix Use a mix of fairly fine bark chips with charcoal and chopped sphagnum moss, or tie plants on to cork or bark slabs and keep roots well misted.

GROWING METHOD

Propagation Plants can be divided after flowering once they have formed a good clump.

Watering Maintain moisture around roots throughout warmer months but allow the mix to dry out between waterings in cool weather. Mist weekly in summer.

Feeding Apply half strength soluble liquid plant food once a month, weekly once buds appear.

Problems There are no specific pest or disease problems but plants can die if they dry out completely or if they are kept sodden.

FLOWERING SEASON

Depends on the species. Most flower through autumn, winter or spring but this can vary.

STANHOPEA

Stanhopea species and hybrids

Native to Nicaragua, Colombia and Venezuela, Stanhopea wardii *has a more pleasant smell than some species of this genus.*

The weird and sinister-looking flowers of Stanhopea tigrina *are a botanical curiosity enjoyed by many growers.*

FEATURES

Epiphytic

Sometimes known as upside-down orchids, stanhopeas must be grown in hanging containers as the flowers emerge from the base of the pseudobulbs and will otherwise be squashed. They push straight through the bottom of the basket and hang down below the foliage. Stanhopeas are evergreen epiphytes from Central and South America. They grow from a fairly large, ribbed pseudobulb and produce large, solitary, dark green leaves. The strange-looking flowers are large, heavy and strongly perfumed. Not everyone finds the perfume pleasant. Flowers are not long lasting but appear in succession. Plants grow rapidly and are very easy to grow into large specimens.

STANHOPEA AT A GLANCE

Easy to grow but can reach 40cm (16in) high. Flowers are large and strongly scented but short lived.

JAN	occasional water
FEB	water and feed
MAR	flowering, water and feed
APR	flowering, water and feed
MAY	flowering, water and feed
JUN	flowering, water and feed
JULY	flowering, water and feed
AUG	flowering, water and feed
SEPT	water and feed
OCT	water and feed
NOV	occasional water
DEC	occasional water

RECOMMENDED VARIETIES

S. graveolans (yellow)
S. oculata (cream)
S. tigrina
S. wardii
S. Assidensis (yellow and red)

Species *Stanhopea tigrina*, with its fleshy yellow flowers blotched dark maroon-red, is the species most often cultivated, although *S. wardii* is also seen. It also has yellow flowers but with plum to purple spots.

CONDITIONS

Climate Prefers a cool, humid climate with a minimum of 10°C (50°F) and tolerates warmer temperatures in summer with shade and high humidity.

Aspect Grows in dappled sunlight in a well-ventilated glasshouse.

Potting mix Line the container with soft coconut fibre or other material so that the stems can push through easily. The mix of coarse bark, alone or with charcoal, must be free draining.

GROWING METHOD

Propagation Divide the pseudobulbs after flowering, but not until the container is full to overflowing. Large specimens are the most rewarding, producing many spikes.

Watering Water freely during warm weather and mist plants if humidity drops. Water only occasionally in winter.

Feeding During the growing season apply weak liquid fertilizer every two weeks.

Problems Can be prone to red-spider-mite or scale insect if not enough humidity is provided. Mist foliage regularly to prevent this.

FLOWERING SEASON

Summer or autumn, depending on species. Remove spent flowers once they have faded.

THELYMITRA

Sun orchid

The glorious flowers of these sun orchids, with their sky blue colour, will not appear in overcast or cloudy weather.

The natural habitat of the sun orchid is heathland and open scrub, and it prefers damp sandy soils.

FEATURES

Terrestrial

Most of these pretty terrestrial orchids are native to damp ground in coastal heaths or open forest in Australia and New Zealand. The species vary greatly in size with plants 15–60cm (6–24in) or more high. Plants grow from small tuberous roots just below the ground. They are herbaceous, becoming dormant in summer. They produce a single, narrow, stem-clasping leaf and from three to a dozen or more flowers. Flowers only open in bright sunny weather as the common name implies. Unlike other orchids, petals and sepals are alike and there is no lip. Flowers are usually blue but may be pink or purple. The yellow *Thelymitra antennifera* is quite difficult to grow. Up to six sun orchids could be grown in a 15cm (6in) pot. Commonly grown in Australasia but not easily obtainable in some countries.

Species Most popular is the spotted sun orchid, *T. ixioides*, with star-shaped flowers, usually blue but sometimes pink, with spotting on the upper petals. Other blue-flowered species are *T. pauciflora* and *T. venosa*.

THELYMITRA AT A GLANCE

Very unusual; not often seen in cultivation. Reaches about 30–40cm (12–16in) in height when in bloom.

JAN	water and feed
FEB	water and feed
MAR	water and feed
APR	water and feed
MAY	rest
JUN	flowering, rest
JULY	flowering, rest
AUG	flowering, rest
SEPT	water and feed
OCT	water and feed
NOV	water and feed
DEC	water and feed

RECOMMENDED VARIETIES

T. ixioides (blue/pink)
T. pauciflora (blue)
T. venosa (blue)

CONDITIONS

Climate Grows in warm to subtropical regions, depending on species.

Aspect Grows best in full sun or very light shade. Flowers will not open if too shaded.

Potting mix Mix must be moisture-retentive but not waterlogged. Try plants in coarse sand with added coconut fibre peat and fine bark.

GROWING METHOD

Propagation Usually grown from seed produced in large quantities after flowers die off in summer.

Watering Water during active growth through winter and spring. In summer water only very occasionally to keep tubers from complete desiccation. Resume regular watering in autumn as new growth appears.

Feeding Use liquid organic fertilizers based on fish or seaweed, or liquid blood and bone, monthly through the active growth period. Use them at half the recommended strength.

Problems No problems if growing conditions are met. Tubers rot if overwatered when dormant.

FLOWERING SEASON

From late winter to late spring, depending on the species.

VANDA

Vanda species and hybrids

Among the best known orchids in cultivation is Vanda *Rothschildiana. It can flower intermittently through the year.*

The unusual colour combination in this Vanda *hybrid, Hawaii, results in a very striking flower.*

FEATURES

Epiphytic

Vandas are lovely epiphytic orchids that bear sprays of long-lasting flowers. However, they do not make good houseplants but are ideal for conservatory or greenhouse cultivation where a high level of humidity can be maintained. The orchids do not have a pseudobulb but grow from strong stems that produce numerous aerial roots. Strap-shaped leaves appear on the upper part of the stems and sprays of flowers emerge from the leaf axils. Flowers are flattish with small lips and are available in a great range of colours; the checkered or tessellated forms are probably the most outstanding. The so-called 'blue orchid', *Vanda coerulea*, from northern India and the hybrid *V.* Rothschildiana, which is violet-blue with darker veining, are among the most famous. Blue is not a common colour in orchids. *V. tricolor* is yellow, patterned in reddish brown with striped purple lips.

Origins These plants are native to tropical regions of the Himalayas and areas from Burma and Thailand to Malaysia. They have been crossed with various other genera for great colour variations.

VANDA AT A GLANCE

Best suited to greenhouse as needs light and humidity. Can flower all year round with flowers up to 10cm (4in) across.

JAN	water
FEB	water
MAR	water and feed
APR	water and feed
MAY	water and feed
JUN	flowering, water and feed
JULY	flowering, water and feed
AUG	flowering, water and feed
SEPT	water and feed
OCT	water and feed
NOV	water
DEC	water

RECOMMENDED VARIETIES

V. coerulescens (miniature pale blue)
V. cristata (miniature green)
V. tricolor (yellow)
V. Rothschildiana (blue)

CONDITIONS

Climate This orchid is frost tender and prefers a temperature range of 12–30°C (54–86°F).

Aspect Provide light shade and high humidity. There must be some air movement and fans may be needed in glasshouses.

Potting mix Pot this orchid in the coarsest mix of bark and even crushed rock. Because of its long aerial roots it does well in a basket. Do not try to confine the roots.

GROWING METHOD

Propagation Not easily propagated but can grow from offsets from the plant base.

Watering Water frequently in summer and mist to maintain humidity around the aerial roots.

Feeding Apply half strength soluble liquid fertilizers about every two weeks during periods of rapid growth in summer.

Problems No specific insect pest or disease problems.

FLOWERING SEASON

Some flower intermittently through the year while others have a definite flowering season. This varies according to species.

Cut flowers Vandas make excellent cut flowers.

VANILLA
Vanilla species

The Vanilla *orchid has a distinctive climbing habit which enables it to grow over the trunk and branches of trees in the rain forest.*

Vanilla flavouring is produced from the seed pods of the flowers of Vanilla planifolia. *The plants are cultivated as a commercial crop.*

FEATURES

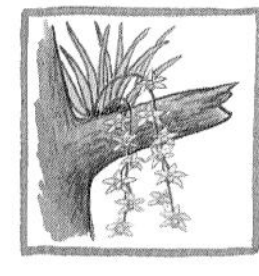

Epiphytic

Everyone knows vanilla flavouring but not many people know that it is from an orchid. The *Vanilla* is the only orchid grown as a commercial crop and although its native homeland is Central America and the West Indies, the plant is grown commercially in many countries, including Madagascar. The species grown is *Vanilla planifolia*, which has a distinctive vine-like habit, enabling it to climb all over trees. The seed pods are pollinated by hand in production nurseries, then allowed to develop before being harvested. This orchid is very much a hot-house plant and would only grow well in a cool climate if kept in a humid warm greenhouse where there is plenty of room for it to climb. Not often grown by amateurs.

VANILLA AT A GLANCE

Needs hot, humid light conditions. Moss pole may be needed to aid straight growth. Flowers are 7cm (2¾in).

JAN	water and feed
FEB	water and feed
MAR	water and feed
APR	water and feed
MAY	water and feed
JUN	flowering, water and feed
JULY	flowering water and feed
AUG	flowering, water and feed
SEPT	water and feed
OCT	water and feed
NOV	water and feed
DEC	water and feed

RECOMMENDED VARIETIES

V. planifolia
V. planifolia var. *variegata*
V. pompona
V. siamensis (leafless)

CONDITIONS

Climate A temperature of at least 18°C (66°F) must be maintained in the winter to make this plant thrive.

Aspect Some greenhouse shading must be provided in the hottest months to avoid over-heating or scorching.

Potting Mix This plant grows as a climber, the base part with roots may be planted in a pot with bark compost. Let it grow up against a fern or moss pole in which it can adhere its roots.

GROWING METHOD

Propagation With the creeping stem habit this plant does not possess pseudobulbs. At the point where each leave joins the stem, it is able to branch and grow another stem. Sections can be removed and grown on in a tray of moist bark and moss. This should not be done too often, though, as the plants needs to grow undisturbed to encourage the flowering.

Watering This orchid needs a great deal of humidity and needs to be regularly sprayed, every day, over its leaves and aerial roots to keep from drying out.

Feeding Apply a weak fertilizer, every ten days, diluted in the spray given.

Problems Should not be attempted to be grown in the home, only in an orchid hothouse.

FLOWERING SEASON

The flowers are produced singly at intervals along the creeping stem at any time of year but mostly summer. Best conditions for flowering are hot, humid, light and when the plant has reached its optimum length.

ZYGOPETALUM

Zygopetalum species and hybrids

This hybrid Zygopetalum *has more intense colour and markings than the species from which it has derived.*

Autumn-flowering, this Zygopetalum mackayi *is known as 'Titanic'. Flowers last well and have a light perfume.*

FEATURES

Epiphytic

Several of these evergreen orchids from tropical America start flowering in autumn, when few other plants are in bloom. Most are epiphytic, occurring in rainforest where they are sometimes found on rocks or among leaf litter. They are generally easy to cultivate and can be teamed with orchids such as cymbidiums that enjoy similar growing conditions. *Zygopetalum* grows from a pseudobulb, producing lance-shaped, mid-green leaves with well-defined veins. The flowers, which are sweetly scented, are an unusual mixture of green, brown and violet-blue. The species most often grown is *Z. mackayi* which flowers from autumn to early winter. *Z. crinitum*, which strongly resembles *Z. mackayi*, has similar flowers in spring and is more compact in size.

ZYGOPETALUM AT A GLANCE

Compact and relatively easy to grow but needs intermediate and shady conditions. Fragant blooms of 5cm (2in) across.

JAN	rest
FEB	rest
MAR	flowering, water and feed
APR	flowering, water and feed
MAY	flowering, water and feed
JUN	flowering, water and feed
JULY	flowering, water and feed
AUG	water and feed
SEPT	water and feed
OCT	rest
NOV	rest
DEC	rest

RECOMMENDED VARIETIES

Z. crinitum
Z. mackayi
Z. Artur Elle
(all green/purple/white)

CONDITIONS

Climate Prefers a moderate climate with a temperature range from 12–30°C (55–86°F). This orchid is generally unsuitable for growing in cooler greenhouses.

Aspect Needs about 50 per cent shade in summer but tolerates full sun in winter.

Potting mix A mix suitable for cymbidiums will suit these orchids. It must be open and free draining to avoid waterlogging.

GROWING METHOD

Propagation Re-pot and divide after flowering about every three years when pots are very crowded, removing leafless backbulbs that can be used for propagation. Roots are rather brittle, so take care when re-potting.

Watering Water freely during spring, summer and early autumn but reduce this once weather cools and growth slows.

Feeding Slow release granular fertilizer or an occasional supplementary feeding of soluble liquid can be given.

Problems *Zygopetalum* is generally easy care and trouble-free. However, large black spots may develop on foliage and leaf tips blacken if conditions are too humid and foliage stays wet in cold conditions.

FLOWERING SEASON

All species flower some time between autumn and spring, the most popular in autumn. Always with new shoots.

Indoors Pots in bloom can be brought into the house where the scent will perfume the room for many weeks.

GROWING BROMELIADS

Bromeliads are attractive plants that are easy to cultivate, both in the greenhouse, as conservatory specimens or as lush, exotic house plants. Their showy but unusual blooms are quite unlike any other flowers, while their foliage, colours and shape hint at their tropical native habitats.

Today many species of bromeliad are threatened in the wild because the forests and woodlands of their natural habitats are rapidly vanishing. There has also been over-collection, and as many species grow very slowly from seed they are not regenerating fast enough to keep up with demand. However, many species are now being cultivated and preserved in botanic gardens and in the collections of both amateur and professional growers, so that home gardeners can continue to grow these interesting plants. Many exciting plants are sold through garden centres, supermarkets and DIY stores – all are well worth buying for the enjoyment they will provide.

LEFT: Exotic foliage and an epiphytic habit make bromeliads rewarding and sometimes challenging to grow. Larger displays for the conservatory or greenhouse can be created by attaching individual plants to old tree branches.

ORIGINAL HABITAT

Most of this large, very diverse group of plants originated in tropical America, with a few species from subtropical America and one species of *Pitcairnia* native to West Africa. Probably the best known of all bromeliads is the pineapple. Its distribution extends from the state of Virginia in the United States south to Chile and Argentina. Bromeliads are most common in rainforests but a few occur naturally in deserts, often dropping their leaves during the driest seasons.

INTRODUCTION TO HORTICULTURE

The first bromeliad introduced to horticulture outside its native habitat was the pineapple, brought to Spain from Guadaloupe in the West Indies by Christopher Columbus on his second voyage to the New World at the end of the fifteenth century. Although it had long been cultivated in the West Indies it caused quite a sensation in Europe.

By the seventeenth century a number of wealthy people were building heated glasshouses in order to be able to cultivate exotic tropical plants such as pineapples, although it was not until the late eighteenth and early nineteenth centuries that heated glasshouse culture became more commonplace. Glasshouses were still, however, the province of the wealthy.

During this period large numbers of bromeliads were introduced into Europe. In the early nineteenth century most went to France and Belgium, where there were the greatest number of enthusiasts and authorities on the subject. By the end of the century, however, collectors from many other European countries were growing and writing about this fascinating group of plants.

This century, with a few notable exceptions, collecting and interest has been most common in the United States, where the Bromeliad Society was established in Florida in the 1950s. Today, however, the cultivation of bromeliads is popular worldwide.

WHERE TO GROW BROMELIADS

In their native habitats, bromeliads thrive in tropical and sub-tropical environments, favoured with heavy rainfall, filtered sunlight and mild or warmish temperatures.

There are a few species that will tolerate cooler conditions although frost, wind and rain during the winter months has an adverse effect on plants – damaging leaves and disfiguring growth. Other desert species will tolerate extreme heat and full sunshine, but generally bromeliads prefer warmth, filtered sunlight and shelter from strong winds. Occasionally plants may be stood outdoors during the summer months in a partially shaded position, but in general they should be confined to the greenhouse or

Vibrant red bracts remain even after the flowering of Guzmania *'Luna' has finished, thus extending the period when this plant is of particular interest.*

'Blue-flowered torch' is the name sometimes applied to Tillandsia lindenii. *It does, in fact, bear deep purple flowers that appear from the pink-flushed floral bracts.*

conservatory and for indoor culture. The occasional trip outside for a shower to remove dust and grime will pay dividends. The wide leaves accumulate dust even in the cleanest of homes, which reduces their ability to absorb sunlight and inhibits healthy growth.

Light is important if you are to enjoy a good show of bracts and flower spikes, and indoors specimen plants can be reluctant to flower. However, the lack of a flower is a small price to pay as the foliage shapes and colours of most bromeliads more than compensate.

Potted plants can be moved around at will, brought into the house or conservatory when they are looking their best, and returned to the greenhouse when they need additional care and attention. In a mixed grouping it is also possible to bring those plants that have flower spikes or are looking particularly attractive to the front of a display, moving less spectacular specimens back for a rest.

One method of display is to make a bromeliad 'tree' so that the plants are seen as epiphytes, looking as they would in their natural environment (see page 78).

FEATURES

In their native habitats some bromeliads are ground dwellers while others grow on rocks or cliff faces, but many are epiphytes using other plants, especially trees, for support although they are not parasites. Most have spirally arranged leaves that channel water into the centre of a rosette, which acts as a water storage tank. In their habitats these 'tanks' provide a home for insects with aquatic larval stages and a breeding place for many species of tree frogs. The leaves also have fine water-absorbing scales although these are rarely obvious.

Leaves may be in various shades of plain green but many have foliage that is attractively patterned in stripes, transverse bands or spots. Silvery grey, deep maroon and bright crimson leaves are also found, along with a range of two-toned or multi-coloured leaves. Some have leaves with smooth edges while others are very spiny or serrated.

One group that displays a vast range of foliage types is *Tillandsia*, which includes many species with a recognizable rosette of leaves as well as the fine, grey, web-like Spanish moss (*T. usneoides*) which grows in long swags draped from tree branches. The leaves are thread-like on stems covered with silver scales.

Bromeliad flowers may be spectacular and showy, held on upright or arching stems, or they may be short, blooming low down in the centre of the rosettes. Some flowers last only a few days while others can remain fresh and attractive over several weeks. The colour range is as extensive as the range of forms. Although some cut well and last a week or two in a vase, they are generally unsuitable as cut flowers. Whole plants in flower are often used instead to create unusual floral arrangements.

CONDITIONS

Containers

When growing bromeliads in pots, it is often necessary to crock the base of the pot with stones or terracotta chips. This provides weight in the base so that the often top-heavy plants do not constantly tip over. Terracotta pots are useful in this regard, as they are heavy enough to provide stability. Many plants are themselves so decorative with their patterned leaves and bright flowers that a simple plain container is the best choice.

Sharing a need for warmth and humidity, greenhouse displays of several different species create an exotic, almost tropical feel. Plants will appreciate shading from strong summer sun and regular misting with rainwater to maintain humid conditions.

Compost

For plants that are happiest clinging to the cleft in a tree or nestled in the leaf litter on a tropical forest floor, the compost used will dictate whether the plants are going to survive or die. The compost must imitate the natural tendency for free drainage coupled with good moisture retention. Nutrients in the wild would come from rotting leaves and dead insects that become trapped in the watery reservoir of the plants. However it is not advisable to feed your plants this way. A compost that contains a controlled release fertilizer can be used, but so long as it is coarse and free draining they will be happy.

Orchid and African violet composts are sold in garden centres and these are ideal – combining the water holding properties of peat or composted bark chips, with the free draining properties of grit or sharp sand. Leafmould would be equally suitable if you can get hold of it.

Planting

Bromeliads are very prone to rotting off if the base of the plant is kept consistently moist. A free draining compost will help, as will plenty of crocks in the bottom of the pot.

When potting it is also important to not plant too deep. The base of the plant should be level with the surface of the compost. Too deep and excess water will not be able to drain away from the central reservoir.

House plants
Light is important for good flowering and foliage colour but even some of the most decorative bromeliads, like the urn plant (*Aechmea fasciata*), will thrive in the home. Bromeliads can be rather large for the average home, although a good range of smaller species and varieties are available. They look particularly striking in simple surroundings, more like a piece of art or sculpture than the traditional potted house plant.

Greenhouse growing
Those with greenhouses have the perfect environment to experiment with a wide range of bromeliads. The warm conditions and bright light will also encourage flowering. Shading in full summer is advisable, as is damping the floor to maintain high levels of humidity. Epiphytic species can be grown on pads of sphagnum moss or old logs and hung from the greenhouse glazing bars for a decorative display. Regular misting will be essential.

Conservatory culture
For the conservatory that is used for plants rather than soft furnishings, bromeliads offer a wide range of colours and forms to play with. They are quite unlike any other house plants but are easy and rewarding to grow. Larger specimens will relish the space a conservatory has to offer, and hanging specimens like Spanish moss (*Tillandsia usneoides*) will add a decorative, almost magical touch. For the more adventurous, a way of displaying a wide range of bromeliad habits is to plant up a 'tree'.

Many bromeliads do well in conditions that suit cymbidium orchids and growers enjoy cultivating these contrasting plant groups together. Both enjoy a free flow of air around them and a degree of humidity at all times.

A bromeliad tree
Tree stumps, especially old tree-fern stumps, make ideal hosts for bromeliads, as do pieces of weathered driftwood. Pieces of driftwood of various sizes make more or less portable 'trees', which can be moved around to suit seasonal conditions or to provide striking decorative effects in different places around the house, greenhouse or conservatory, perhaps for a special event.

To attach plants to the tree, wrap the roots in sphagnum moss or peat and fix them on firmly with upholstery webbing, nylon fishing line or old stockings. Small specimens may be attached using a little PVA glue, although you should not let the glue smother the roots.

The roots of small plants may be passed through the holes in rounds or squares of 1–2cm (½–1in) plastic mesh and a pad of sphagnum moss placed underneath them before the mesh is stapled or tied to the tree. The type of plastic mesh sold to keep leaves out of guttering is ideal. An artificial log can also be made from this mesh by forming a cylinder and packing it with chopped bark and moss. Push the plant roots through into the growing medium and tie them on with plastic-coated wire or fishing line. Once plants are established the ties can be removed. These cylinders are light and can be suspended from greenhouse staging, glazing bars, conservatory support bars or hung in partially-shaded windows.

To grow bromeliads successfully indoors there must be plenty of light or the rich colour of the foliage will not be maintained. Plants should be positioned near a sunny window but not close to the glass where the leaves may be scorched. If there is not enough light it will be obvious after a time that leaves are becoming pale and drawn. Extra light may need to be provided artificially.

Plants also need a fairly humid atmosphere, so if you have central heating or the room gets warm, sit plant pots on a bed of pebbles or gravel in a saucer of water. The pot base must be above the water level so that the potting mix is not constantly sodden. Spray-misting the plants with water is also beneficial. This can be done daily in summer but much less often in winter. Foliage should be wiped down with a damp cloth from time to time to remove dust or taken into a gently running shower for a good rinse. If possible take plants outside and gently hose them down.

GROWING METHOD

Watering
Bromeliads growing on logs or moss pads are best watered by spray misting. If you only have a few plants this can be done with a hand sprayer but larger collections will need hosing with a fine spray. Those with a vase form are best

BROMELIAD TREE

watered from above, allowing the water to fill the central cup or reservoir. As this overflows it will give roots the moisture they need. In the warmer months when plants are in active growth they may need watering two or three times a week, with misting on the other days. In winter watering once a week should be enough. If the water in your area is 'hard' (has a high mineral concentration) you should collect rain water to use on your bromeliads.

Feeding
Opinions vary on the value of feeding these plants. In nature leaves and other debris fall into the vases of plants, slowly decaying and providing nutrients to the plant. If you do fertilize, make sure that it is in very weak concentrations, not too often and never in winter. Some professional growers prefer to use a little slow release fertilizer around the root zone of the plant while others use well-diluted water soluble fertilizers as a spray. If using a spray, use only a quarter to half the recommended dilution rate given for other foliage or indoor plants, and apply it only after the plant has been watered and when it is not in the sun or on very hot days. It is best not to pour even dilute fertilizer into the vase of the plant as this may scorch the foliage.

PROPAGATION

Plants that have bloomed will slowly die off over the next year or two. However, they will replace themselves by producing offsets, often called 'pups', generally from the side of the base of the plant. It is best to wait until the offset has a firm base and is at least one-third the size of the parent plant before removing it. Cut off the offset as close as possible to the mother plant using a sharp knife or secateurs. This is best done some time between spring and mid-autumn. A few plants produce pups higher up on the plant and these are easily detached by hand.

Any discoloured or brown leaves should be removed before potting up, and long, woody running stems (stolons) should be shortened. Offsets should be potted up into small pots containing a coarse seed and cutting compost or two parts sharp sand to one part peat or composted bark. Some growers like to dust the base of the offset with a fungicide as a disease preventive. Don't push the offsets in too deep or they may rot. Spray mist daily, unless there is a cold spell, and roots should form on the offset in a matter of weeks. The best time to propagate by division of offsets is during the growing season, which is from spring through to mid-autumn.

The striking cream and green striped foliage of this Guzmania *cultivar makes this bromeliad showy and decorative all year round. The red floral display is an added bonus.*

WHAT CAN GO WRONG?

Adequate spacing of plants, good ventilation and good cultural practice should minimize problems. Remove any dead or decaying material to keep plants clean and looking good. If you have a sick plant, isolate it from the rest of the collection and wash your hands and disinfect any tools used before handling healthy plants. If you have to use chemicals for pest or disease control, do not spray the buds or flowers as they may become distorted. Fungicidal powders may also be used.

Pests
- Scale insects. These small sap suckers appear as soft brown spots that can be easily pushed off with a finger nail. Wipe off scales with a damp cloth or spray with a suitable insecticide at half the recommended strength. Never use oil-based formulations on bromeliads as this smothers the natural breathing pores and the plant will die of suffocation.
- Mealybugs. These soft white insects resemble small sticky pieces of cotton wool. They are usually found clustered in the leaf axils and so may be difficult to reach. Wipe them off if possible or use a suitable insecticide. They are usually only a problem if the bromeliads are grown in crowded or badly ventilated conditions.

General disorders
- Lowest leaves brown at the base. This may be caused by heavy, poorly drained mix, overwatering or deep planting.
- Inner leaves rolled or stuck together. The air may be too dry, the plant may need misting or the central vase may have completely dried out.
- Brown leaf tips. These may result from poor drainage or from overwatering, or indicate that the atmosphere is dry.
- Brown patches on leaves. These may be caused by strong sun or very high light intensity, overwatering or poor drainage.
- Bottom leaves straw to light brown in colour. This is usually a sign of natural ageing as the older leaves die off.

AECHMEA

Aechmea species and cultivars

The floral bracts of 'Ensign', a cultiver of Aechmea orlandia, *are made up of tightly overlapping, scarlet segments.*

Long flowering and adaptable, Aechmea weilbachii *makes a good choice of plant for fairly warm conditions.*

FEATURES

Epiphytic

These are possibly the most widely grown and best known of all the bromeliads. Generally very easy to cultivate, they can be container grown or attached to a log 'tree'. Although aechmeas originate from Central and South America, most tolerate temperatures down to about 5°C (41°F) making them suitable for cooler conservatories. Species vary in size from a petite 15cm (6in) to over 60cm (24in), although this size would be rare in cultivation. Larger plants can become top heavy so a terracotta or earthenware pot is advisable for extra stability.

Foliage There is an amazing diversity of form and foliage colour in these bromeliads. Leaves are in a rosette, forming a vase shape with an open cup in the centre. They may be completely shiny green, or green above and burgundy beneath, deep burgundy on both sides or streaked, banded or spotted with silver, or entirely silvery black. There are also variegated forms with either yellow-gold or pink variegations on a green background. Many are sharply spined along the leaf margins (needing careful handling) while others are almost smooth edged.

Flowers Many aechmeas have long-lasting flowers that make them popular for indoor decoration. Although the true flowers are often quite small they are enclosed within showy bracts that come in a great range of colours. The inflorescence or flowering spike may be red, blue, yellow, purple, pink, orange or white and lasts for months.

Fruits Berry-like fruits follow the flowers and often persist on the plant over a long period.

A. fasciata One of the most popular species is *Aechmea fasciata* or 'Urn' plant, which has grey-silver spined leaves cross-banded in silver. The whole leaf surface is densely covered with silvery scales, giving a powdery effect. The flower spike is a large, showy pyramid of pink bracts enclosing the blue flowers, which age to red.

Foster's Favorite An attractive dwarf variety is *A.* 'Foster's Favorite', which has deep burgundy foliage

AECHMEA AT A GLANCE

A tolerant plant. Leaves are strap-like and vary in colour. Plants produce floral bracts after three years.

JAN	grow on, reduce watering
FEB	grow on, reduce watering
MAR	re-pot, feed
APR	remove and pot on offsets
MAY	remove offsets, mist foliage
JUN	bracts, flowers, mist and water
JULY	flowers, mist and water
AUG	/
SEPT	/
OCT	flowers, fruit; keep frost free
NOV	flowers, reduce watering
DEC	flowers, reduce watering

RECOMMENDED VARIETIES

A. chantinii
A. fasciata
A. fulgens discolor

Aechmea fasciata *is one of the easiest bromeliads to grow and produces a magnificent pink flowerspike that will last for months.*

With purple undersides to the leaves, Aechmea fulgens discolor *also produces a stunning coral red flowerspike that is covered in berries.*

and a pendulous flowering spike of deep blue flowers. These may be followed by red berries. Both these forms are tolerant of a wide range of conditions.

A. chantinii *A. chantinii* is variable both in size and foliage colour. It is sometimes called Amazonian zebra plant because of its green to almost black foliage, which is heavily barred. The long-branched flowering spike is generally red or orange with flowers being red or yellow. Tends to be more cold sensitive than some.

Other species *A. fulgens discolor* has attractive foliage and is commonly called the 'Coral berry' aechmea. This species is hard to beat with its green strap-shaped leaves that are a deep purple beneath. A spike of purple flowers will turn into decorative coral red berries.

CONDITIONS

Position Needs a frost-free climate. The plants do best in warm to hot, humid conditions with a cooler spell in winter. Morning sun, filtered sunlight or shade seems to suit this plant group. Most aechmeas like sheltered situations, preferably with overhead shading. Although some species have origins in harsh environments it is best to give them all some shelter in non-extreme conditions.

Potting mix Plants will thrive in a very coarse, open, soilless compost. Water must be able to drain straight through. Take care not to plant too deep to avoid rotting.

GROWING METHOD

Propagation Start new plants by removing the offsets or pups from the parent plant once the offsets have reached about one-third of the size of the parent. Cut off and pot separately.

Watering The cup at the centre of the rosette must be kept filled with water. Plants probably need watering twice weekly or more in summer and every week or two in cold weather. Be guided by the weather and feel how moist the compost is. Mounted plants need spray watering daily in summer but much less often in winter.

Feeding Apply slow release granules to the compost in spring. Mounted specimens may be given a foliar spray of liquid plant food at about one-third the recommended strength. Over feeding will not encourage more vigourous growth – it will scorch the leaves and roots.

Problems There are no specific problems for this group if given reasonably good cultural and environmental conditions.

FLOWERING SEASON

Flowering times are variable but many bloom in late summer and autumn, and many of them continue into winter.

ANANAS

Ananas species

The pineapple with its familiar crown of stiff leaves develops in the centre of the plant. This one is almost mature.

The forms of Ananas *with variegated leaves are very attractive year round, even without the flowers.*

FEATURES

Terrestrial

Pineapple, *Ananas comosus*, is one of several species that make up this terrestrial bromeliad genus. All originate in tropical America. They have a rosette of very stiff, spiny leaves and produce purple-blue flowers with red bracts on a stem rising from the centre of the plant. After the flowers fade the fruit is formed. *A. bracteatus* is grown for its showy flowers, which are followed by bright red mini pineapples. The variety *striatus* has leaves edged and striped cream to white. Unfortunately, to produce pineapples *A. comosus* must be grown in the right conditions. The form with cream striped leaves is the most popular. In bright light variegations may turn pinkish. Take care when siting these plants as the foliage spines are sharp.

ANANAS AT A GLANCE

To produce pineapples grow in a hot conservatory. Flowers are purple-blue with red bracts; fruit forms after flowers.

JAN	reduce water, move to 10°C (50°F)
FEB	water every two weeks
MAR	remove and pot on offsets
APR	remove and pot on offsets
MAY	feed and light
JUN	flowering, water three times weekly
JULY	flowering
AUG	water three times weekly
SEPT	water every two weeks
OCT	water every two weeks, keep frost free
NOV	water every two weeks
DEC	water every two weeks

RECOMMENDED VARIETIES

A. comosus
A. comosus striatus

CONDITIONS

Position Needs a frost-free climate with a winter temperature above 10°C (50°F). Needs full sun or very bright light to flower and fruit. Very bright light also brings out the best colour of variegated forms.

Potting mix All growing media must be well drained. Use coarse bark or peat-based mix and a heavy pot for additional stability.

GROWING METHOD

Propagation Grows from suckers or offsets from the base of the plant or from the tuft of leaves on top of the fruit. Peel off the lower basal leaves to reveal a stub and leave the stub in a dry, airy place to dry before planting it sometime from spring to autumn.

Watering In summer water two or three times a week. In winter check before watering, which may be needed only every week or two.

Feeding Give slow release fertilizer in spring and early summer if desired.

Problems No specific problems are known for home growers but base and stem will rot if plants are too wet.

FLOWERING SEASON

Flowers appear from late spring to summer, depending on the season.

Fruit Fruit may take two years or more to mature, especially in cooler conditions, but the foliage makes up for this.

BILLBERGIA

Billbergia species

Bright red, overlapping bracts almost conceal the small flowers of this showy Billbergia *hybrid.*

This pendulous inflorescence reveals a mass of small blue-green flowers emerging from pink bracts.

FEATURES

Epiphytic

Terrestrial

One of the most easily grown of all bromeliads, billbergias are widely grown as house plants and are suitable for colder rooms in the house. They adapt to a wide variety of conditions, making them a good choice for the beginner. Leaves are rather stiff and form tall, tubular rosettes. Foliage is spiny and may be mottled, banded or variegated in colours from mid-green to blue- or grey-green. Flower spikes often arch or droop. Flowers are generally not long lasting but some species flower on and off all year. Bracts are often pink or red with green or blue flower petals. Queen's tears, *Billbergia nutans*, is probably the most common. It has narrow, grey-green leaves to 30cm (12in) with blue and green flowers and pink bracts. It has been widely used in hybridizing.

Other species *B.* x *windii* is a much larger leaved species producing 45cm (18in) flower spikes over the grey-green leaves. Also, look out for *B. zebrina* and *B. pyramidalis* – both larger, more exotic species.

BILLBERGIA AT A GLANCE

Suitable for mounting on a log or in a pot. Flowers on and off all year but not long lasting. Pink bracts are spectacular.

JAN	flowering, water
FEB	water
MAR	/
APR	feed
MAY	flowering, mist, repot
JUN	water and mist
JULY	water and mist
AUG	buy plant
SEPT	water
OCT	/
NOV	remove offsets
DEC	flowering

RECOMMENDED VARIETIES

B. decora
B. nutans
B. nutans 'Variegata'
B. pyramidalis
B. x *windii*

CONDITIONS

Position Needs frost-free conditions with a minimum temperature above 5°C (41°F). Most species do best in fully sunny locations but need shade from the hottest summer sun, which tends to scorch leaf tips.

Potting mix Any open, free-draining mix is suitable. Many experienced growers consider this plant does best without enriched soil conditions.

GROWING METHOD

Propagation Grows fairly easily from divisions or offsets of older plants taken during the winter months.

Watering Don't water too frequently but keep the cup filled with water. Spray misting to maintain a humid atmosphere around the plant is an ideal way to maintain good growth.

Feeding Some growers advocate regular liquid feeding through the growing season, others prefer not to give supplementary feeding.

Problems There are generally no problems.

FLOWERING SEASON

Flowering times vary according to species and growing conditions.

CRYPTANTHUS

Earth stars

Best viewed from above, earth stars are small, stemless and low growing. They can do well growing in terrariums.

Leaf margins that are soft and wavy and variable pink striping are typical foliage characteristics.

FEATURES

Terrestrial

These terrestrial bromeliads have compact, attractive foliage in a flat, star-like shape and make excellent pot plants. All species have insignificant white flowers in the centre of the plant. Leaves usually have wavy margins and come in a great range of colours—green, cream, brown, pink and silver—and patterns, and they form low, spreading rosettes 5–30cm (2–12in) across. *Cryptanthus bromelioides* var. *tricolor* is known as the rainbow star for its striped cream and green foliage flushed with rose pink. It is the hardest to grow and can reach 30cm (12in) across. *C. zonatus* has wavy green foliage banded in silver, grey or brown and can be 40cm (15in) across. *C. bivittatus* grows just 6–8cm (3–4in) across, with pink and green striped leaves.

Species The green earth star, *C. acaulis*, which may grow up to 30cm (12in) across, is better known for its many cultivars which have leaves flushed with red or pink.

CRYPTANTHUS AT A GLANCE

Ideal in a pot in a warm, humid room. Varieties have rosettes of coloured foliage. Not good for flowers.

JAN	water
FEB	water
MAR	good time to buy, re-pot
APR	mist, feed, remove offsets
MAY	flowering
JUN	flowering, mist, feed
JULY	flowering
AUG	flowering
SEPT	/
OCT	keep frost free
NOV	keep frost free and soil moist
DEC	keep frost free and soil moist

RECOMMENDED VARIETIES

C. bivittatus
C. bivittatus 'Pink Star'
C. bivittatus 'Roseus Pictus'
C. bromelioides
C. x *roseus* 'Le Rey'
C. x *roseus* 'Marian Oppenheimer'

CONDITIONS

Position Grows best in humid, warm to hot conditions but needs to be kept dryer and frost-free during the winter months. Prefers bright, diffused light for best foliage colour but species vary in their needs and some tolerate shade. None likes direct sun.

Potting mix Use an open and free-draining mix with some organic matter to prevent total drying out. African violet composts are ideal.

GROWING METHOD

Propagation Detach offsets from between the leaves, stolons (running stems) or from the base of the plant when one quarter to one third the size of the parent and place firmly into the propagating mix.

Watering The mix should be kept slightly damp at all times although in winter water just often enough to maintain a damp feel to the mix.

Feeding Apply slow release fertilizer in spring and midsummer or very weak liquid fertilizer once a month through the growing season.

Problems Overwatering in cool conditions will rapidly cause rotting.

FLOWERING SEASON

The not very noteworthy flowers appear in late spring or summer.

GUZMANIA

Guzmania species

Shining foliage in pink and green forms a most decorative rosette in this variegated Guzmania.

Gorgeous red flower bracts tucked into green leaves make this G. lingulata *'Empire' look like a decorated Christmas tree.*

FEATURES

Epiphytic

This is a large group of mainly epiphytic bromeliads with a few terrestrial species. They are grown for their lovely spreading rosettes of satiny, smooth-edged foliage, as well as for their striking flowering stems. They have been widely hybridized with *vrieseas* to produce stunning cultivars. Mature plants may be from up to 1m (3¼ft) wide when fully mature. *Guzmania lingulata* is a handsome species with shiny, mid-green leaves, and a rich, bright red inflorescence. Leaves can be up to 45cm (18in) long. *G lingulata minor*, the scarlet star, is much smaller with leaves just 13cm (5in) long. Named varieties include 'Exodus', 'Empire', 'Cherry' and 'Gran Prix'.

GUZMANIA AT A GLANCE

Ideal for conservatory as pot plant or mounted. Grown for rosettes of spineless foliage and bright red flowering stem.

JAN	water
FEB	keep warm
MAR	keep warm
APR	repot
MAY	remove suckers and offsets
JUN	flowering, mist, feed
JULY	flowering, mist, water
AUG	mist
SEPT	reduce misting
OCT	keep frost free
NOV	keep frost free
DEC	keep frost free

RECOMMENDED VARIETIES

G. dissitiflora
G. lindenii
G. lingulata
G. monostachya
G. 'Amaranth'
G. 'Cherry'
G. lingulata 'Empire'
G. 'Exodus'
G. 'Gran Prix'

Leaves Leaves may be plain glossy green, cross-banded in contrasting colours or finely patterned with stripes. At flowering time the central leaves may colour, adding to the brilliant colour display.

CONDITIONS

Position Grows happily in a warm, frost-free greenhouse or conservatory or on a bright windowsill in the home. Prefers bright filtered light away from draughts.

Potting mix Needs a free-draining mix able to retain some moisture or use ready-made orchid compost. Use a pot that is just slightly larger than the root ball. Terracotta pots will give larger plants more stability.

GROWING METHOD

Propagation Grows from offsets or suckers that develop around the stem of the parent plant. Plant out from spring to autumn.

Watering Mist daily in summer. Keep water in the cup at all times and water the potting mix twice weekly in summer and just occasionally in winter as necessary.

Feeding Use weak liquid plant foods during periods of rapid growth. Do not feed too early in spring as it can scorch the leaves and roots.

Problems No specific problems provided suitable cultural conditions are given.

FLOWERING SEASON

The showy flowers are long lasting on the plant – perhaps up to two months. Most species and varieties flower during summer and last well into autumn.

NEOREGELIA

Neoregelia species

This bromeliad with its fiery red centre is aptly named 'Inferno'. It is an outstanding example of Neoregelia.

The leaves of this variegated Neoregelia *are outlined in cream, giving it prominence among darker Aechmea hybrids.*

FEATURES

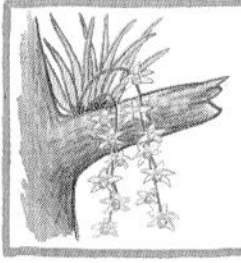

Epiphytic

Terrestrial

Often called heart of flame or blushing bromeliads, neoregelias are very popular for their ease of culture and their dazzling variety. In nature they grow as epiphytes on trees or as terrestrials. Species vary from tiny plants not more than 5cm (2in) wide to those spreading to over 1m (3¼ft). They can be grown as houseplants or as epiphytes attached to logs in the conservatory (see page 78). However, to enjoy them at their best they should be sited low where their beauty can best be appreciated. The group has been widely hybridized, resulting in some truly outstanding cultivars.

NEOREGELIA AT A GLANCE

Varieties vary in size and flowers do not last long. Similar to guzmanias but broader leaves. Keep frost free.

JAN	water
FEB	water
MAR	re-pot if needs it
APR	remove offset
MAY	move away from sunlight
JUN	mist and water
JULY	flowering, water
AUG	buy plant
SEPT	reduce water
OCT	water crown
NOV	keep frost free
DEC	keep frost free

RECOMMENDED VARIETIES

N. carolinae
N. carolinae marechalii
N. carolinae tricolor
N. spectabilis

Flowering Most varieties produce a startling colour change in the centre of the plant at flowering time, the colour remaining long after flowering has ceased. This colour is mostly red, hence the common name 'Blushing Bromeliad'. This group lacks the tall, showy flowering spikes of other genera as the flowers – purple, blue or white – form in the centre of the leaf rosette.

Foliage Leaf rosettes are wide and spreading, the foliage shiny with serrated margins. Leaves may be plain green, red, burgundy or patterned with stripes, bands or spots, or even marbled.

N. carolinae *Neoregelia carolinae* is undoubtedly the most commonly grown and numerous lovely hybrids have originated from this species. The straight species forms a compact rosette with leaves about 25cm (10in) long. The colour of the centre at flowering varies through shades of crimson to cerise and the flowers are deep violet. *N.c. tricolor* has foliage that is cream and green striped. This takes on a pinky red flush as flowering begins and the centre of the plant turns crimson. Other varieties of this species include those with cream or white margined leaves. *N. carolinae marechalii* is another fine species but without the cream stripes. Leaves are plain olive green but flushed with crimson at the base during flowering.

N. spectabilis The fingernail plant, *N. spectabilis*, has red-tipped olive green leaves banded grey on the undersides. A hardy species, it is best grown in bright light, where the undersides of the leaves take on a rosy pink colour. Place high

Small flowers are forming in the vase of leaves on this Neoregelia. *The orange-red shading on the leaves is an added bonus.*

This bromeliad with dark foliage is named 'Hot Gossip'. Speckled bronze-green leaves are margined in deep pinky red.

up on a shelf so that the grey-barred, pink foliage is seen to advantage.

Other species Another species used in hybridizing is *N. fosteriana* which features burgundy foliage. *N. marmorata* has wide leaves growing about 30cm (12in) long. They are marbled in red on both sides and have red tips. *N. eleutheropetala* has sharply spined mid-green leaves that turn purple-brown at the centre. The inflorescence mixes white flowers and purple-tipped bracts.

CONDITIONS

Position These plants must be grown in frost-free conditions with a minimum temperature of 10°C (50°F). A cool greenhouse, conservatory or light room are perfect. Most neoregelias grow well in filtered or dappled sunlight. Where summers are very hot with long hours of sunshine, greenhouse shading may be needed. Indoors these plants will thrive in bright light but not direct sun through a window. For shady spots with no direct sun, the plain green leaved varieties will do best.

Potting mix The compost must be free draining and coarse to allow air to the roots. A mix of bark and gravel or coarse sand is suitable, with added charcoal if this is available. Don't overpot as roots may not utilize all the mix and watering becomes a problem. A pot large enough for a year's growth is ideal – stones or large pebbles can be put in the base to prevent the plant toppling over. Keep the leaf bases just above soil level.

GROWING METHOD

Propagation Detach offsets from the parent plant once they are a good size. New roots form more rapidly if the offset is potted into a seed-raising compost mix or a mix of sand and peat or peat substitute, whichever you prefer.

Watering Water should be kept in the cup at all times. Water about twice a week in summer, with daily misting unless the atmosphere is extremely humid. In winter water only occasionally. When watering, flood the central cup so that stagnant water is changed to avoid problems of rot.

Feeding In the house, greenhouse or conservatory, plants can be given slow release fertilizer when active growth resumes in mid-spring and again in early to midsummer. Many growers believe plants grown without fertilizer produce more vibrant colours. Feed once a month in summer or when the plant is actively growing.

Problems There are no specific problems if cultural conditions are suitable. Apart from rots, usually caused by overwatering in cool weather, dying leaf tips are a sign of trouble. This symptom could be caused by cold, by dry, hot conditions, by drought or by frequent overwatering.

FLOWERING SEASON

Flowers are short lived. They do not usually rise above the rim of the cup. Most flower during late spring or summer.

PITCAIRNIA

Pitcairnia species

Reminiscent of a burning torch, the flame coloured inflorescence of Pitcairnia smithiana *is very striking.*

Terrestrial species of Pitcairnia *are growing here in quite poor soil, but it is open and drains well.*

FEATURES

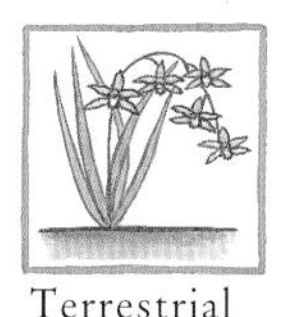

Terrestrial

Epiphytic

This large group of bromeliads contains the only non-American species of bromeliad, *Pitcairnia feliciana*, native to West Africa. Most pitcairnias are terrestrials but some are epiphytes or rock dwellers; many are found in damp, shady habitats. Some lose their leaves during winter dormancy, which coincides with the dry season in their native regions. The fairly narrow leaves may be smooth or spiny edged and the plants can give the effect of grass hummocks in their native habitats. Plants range from 30cm–2m (12in–6½ft) tall. The flower spikes are tall and slim with individual flowers lasting for one day, but flowers are numerous so the display can continue for several weeks.

PITCAIRNIA AT A GLANCE

Keep in humid room indoors. Ideal for planting on log or in pot. Flowers regularly but last one day only.

JAN	keep dry
FEB	flowering, dry
MAR	remove offsets and suckers
APR	remove offsets and suckers
MAY	water twice a week
JUN	feed with slow release fertilizer
JULY	keep out of sun
AUG	flowering
SEPT	/
OCT	reduce water
NOV	reduce water
DEC	reduce water

RECOMMENDED VARIETIES

P. feliciana
P. flammea
P. smithiana

Flowers are usually red or yellow but may be shades of white or cream.

Species — *P. flammea* is an easy-care evergreen species with dark green leaves and coral red flowers, usually in spring.

CONDITIONS

Position — All need frost-free conditions. Some species prefer fairly shady situations while others require more bright light and warmth. Warm summers and cool winter conditions are required for healthy growth.

Potting mix — The mix should be coarse, open and well drained. Some species prefer a growing medium that retains a small amount of moisture at all times.

GROWING METHOD

Propagation — Most species grow readily from offsets or suckers taken from the parent plants during spring to autumn.

Watering — Water about twice a week in summer. Some species need occasional winter watering while others need to be kept quite dry.

Feeding — Apply a little slow release fertilizer in spring or give an occasional liquid feed during the active growing period. Foliar feeds are best applied after the plant has been watered to avoid scorching the leaves and roots.

Problems — No specific problems. Excess winter watering will kill plants, especially those from habitats with distinct seasons.

FLOWERING SEASON

Flowering times vary but many pitcairnias flower during winter or spring.

PUYA

Puya species

The grey leaves with their spiny edges and the whole shape of this Puya *species make it look like a giant starfish.*

A heavy pink stem supports the large inflorescence of Puya venusta, *which here is part of a large collection of the species.*

FEATURES

Terrestrial

This group of terrestrial bromeliads contains the largest species known, *Puya raimondii* from Peru and Bolivia, which is capable of growing to 3–4m (9–12ft) high. This very slow-growing plant takes up to 100 years to produce its first flower spike, which contains thousands of individual flowers. Puyas are mostly terrestrial, although some are rock dwellers, and most come from inhospitable habitats in the Andes. In nature most are pollinated by humming birds or starlings. Some come from cold, damp, windswept regions, others from dry grasslands where intense sunlight, heat and drought are balanced by heavy winter frosts. Many team well with succulents that require similar conditions.

PUYA AT A GLANCE

Varieties vary in size. Flowers any time of year and last for a long time. Can go outside in summer. Keep frost free.

JAN	keep dry
FEB	keep dry
MAR	water
APR	remove offsets
MAY	repot
JUN	water
JULY	water, take outside
AUG	/
SEPT	bring in
OCT	reduce water and remove offsets
NOV	keep dry
DEC	keep dry

RECOMMENDED VARIETIES

P. alpestris
P. berteroniana
P. chilensis
P. coerulea
P. mirabilis
P. venusta

Appearance Most are large, from 1m (3¼ft) upwards, and grow in clumps so that ample space is needed. The heavily spined leaves may be green or grey and silver and are a decorative feature. They form dense rosettes from which tall spikes of flowers appear. Flowers are green, violet, blue or white, often with colourful contrasting bracts.

Species *P. venusta* grows to about 1.3m (4¼ft), producing eye-catching purple flowers on a tall, rose-pink stem and bracts. *P. berteroniana*, over 1m (3¼ft) high, has metallic greenish blue flowers.

CONDITIONS

Position Many tolerate cold winters if kept dry. Most endure extremes of climate with very high daytime temperatures and freezing nights. Grows best in the large conservatory or greenhouse border.

Potting mix The growing medium must be coarse and well drained. A mix of coarse sand and crushed rock with added peat or a peat substitute would be suitable.

GROWING METHOD

Propagation Remove offsets from spring to autumn.

Watering Water regularly to establish plants but once established they need only occasional deep watering while in active growth.

Feeding Little or no fertilizer is needed.

Problems Generally trouble-free and easy to grow.

FLOWERING SEASON

Flowering times vary with species and district. Most have long-lasting blooms.

TILLANDSIA
Tillandsia species

Fairly common in cultivation, the attractive Tillandsia fasciculata *features an unusual inflorescence of three or four stems.*

Starbursts of silver grey seemingly suspended in mid-air, these tillandsias are growing in baskets along with Spanish moss.

FEATURES

Epiphytic

Tillandsias are mostly epiphytes with very poorly developed root systems, and some absorb water and nutrients through their foliage. Habitats vary from sea level to high altitudes and even the desert. One of the best known is Spanish moss, *Tillandsia usneoides,* with thread-like leaves on long silvery stems. Most species form rosettes of green, grey or reddish foliage, and those from arid regions have silver scales. Soft green-leaved species are generally native to humid forests and adapt well to pot culture, while many from arid regions are more easily grown on bromeliad 'trees' or moss pads. Flowers are tubular and may be violet, white, pink, red, yellow, blue or green.

TILLANDSIA AT A GLANCE

Easy to look after; grow on a log. Flowers are varied and can appear in almost any month of the year.

JAN	mist twice weekly
FEB	mist twice weekly
MAR	mist twice weekly, remove offsets
APR	keep at 16°C (61°F); mist twice daily
MAY	flowering, mist
JUN	flowering, mist
JULY	flowering, mist
AUG	flowering, mist
SEPT	mist
OCT	mist twice weekly, remove offsets
NOV	mist twice weekly
DEC	mist twice weekly

RECOMMENDED VARIETIES

T. abdita
T. argentea
T. bulbosa
T. butzii
T. cyanea
T. usneoides

CONDITIONS

Position Most species need frost-free conditions – a cool greenhouse or conservatory is ideal. Green-leaved species need filtered sunlight year round while the grey- or silver-leaved varieties can be grown in full or partial sun. With a mixed collection it may be advisable to provide filtered sunlight, especially if the humidity is low.

Potting mix The mix must be very open and well drained. Use fairly coarse composted bark or a special orchid mix, sold at garden centres or DIY stores. Driftwood, logs or cork slabs are ideal for mounting plants.

GROWING METHOD

Propagation Grows from offsets produced sometime during spring to autumn. (A few species produce offsets between the leaf axils: these may be difficult to remove without damage.) When they have been cut from the parent plant, allow the bases to dry for a few days before fixing them in their permanent positions with a little PVA glue.

Watering Water or mist plants daily during hot weather. Mounted specimens can suffer if not moistened daily. In cool weather mist several mornings a week.

Feeding Not necessary although a very weak liquid feed during the warmest months of the year may encourage better growth.

Problems No specific problems are known.

FLOWERING SEASON

Flower form and colour are variable. Most flower in late spring or summer.

VRIESEA

Vriesea species

The leaves of Vriesea gigantea *are finely checkered or tessellated. This broad, spreading species can grow to 1m (3¼ft) across.*

Parrot feather is a name sometimes given to Vriesea psittacina. *This variable species may grow 40–60cm (16–24in) wide.*

FEATURES

Epiphytic

Terrestrial

Vrieseas are very adaptable, tolerating conditions in the home, conservatory or greenhouse. Most are epiphytes growing on trees in forests but some larger species are terrestrials. Some species in the wild are pollinated by nocturnal insects attracted to the scented flowers. Few of these larger terrestrials are grown outside specialist collections. Leaves are spineless and may be plain glossy green or attractively banded, spotted or variegated. They form neat rosettes. Many species have striking bracts. The true flowers are usually yellow, green or white but the bracts may be red or purple, yellow or green. Plants may be 15–20cm (6–8in) high or reach over 4m (13ft). Some are very wide spreading.

VRIESEA AT A GLANCE

Ideal for mounting or in a pot inside. The scented flowers are long lasting and appear any time of year.

JAN	keep air moist
FEB	water centre of plant
MAR	feed
APR	remove offsets, repot
MAY	water and mist
JUN	mist; keep humid
JULY	mist, needs high temperatures
AUG	mist, water centre of plant
SEPT	mist
OCT	remove offsets
NOV	keep frost free and water centre of plant
DEC	/

RECOMMENDED VARIETIES

V. hieroglyphica
V. carinata
V. x *poelmanii*
V. x *polonia*
V. saundersii

CONDITIONS

Climate Copes with very high temperatures if not direct sun; does best in humid conditions. Many species tolerate low, frost-free temperatures. Most prefer bright, filtered light and good air circulation, again similar to the conditions favoured by orchids. A group of these plants will create a more humid microclimate.

Potting mix Need good drainage and aeration. Use coarse bark, sand, gravel and charcoal as the base, with leaf mould, well-decayed compost or even polystyrene granules added.

GROWING METHOD

Propagation Grow from offsets produced at the base of the plant during spring to autumn. In spreading species they will be under the foliage: once they are sufficiently advanced remove them before they distort the foliage.

Watering Keep the cup in the centre of the rosette filled. In summer, water two or three times a week, spray misting on the other days or if humidity is low. In the conservatory or greenhouse, damp down the floor on hot days. In winter, water only every couple of weeks but maintain atmospheric humidity.

Feeding Apply slow release granular fertilizer in spring and midsummer, or use soluble liquid foods monthly at half the recommended strength. Ensure fertilizer does not touch foliage. Feed only in the warmer months.

Problem No specific problems are encountered.

FLOWERING SEASON

Species flower at different times of the year. Most have long-lasting flowers.

Index

This edition published in 2002 by Murdoch Books UK Ltd
First published by Merehurst Ltd 2000
Merehurst is an imprint of Murdoch books UK Ltd
Ferry House, 51–57 Lacy Road, Putney, London, SW15 1PR

ISBN 1-85391-862 8

COMMISSIONING EDITOR: Helen Griffin
SERIES EDITOR: Graham Strong
TEXT: Sara Rittershausen and Mike Pilcher
EDITORS: Rowena de Clermont-Tonnerre and Christine Eslick
DESIGN: Maggie Aldred and Michele Lichtenburger
ILLUSTRATIONS: Sonya Naumov
Group CEO: Juliet Rogers

Printed by Tien Wah Press in Singapore

PHOTOGRAPHS
All photographs by Lorna Rose © Murdoch Books, except those by:
Derek Cranch © Merehurst Limited: front cover, pp 1, 2, 4, 10 (all images), 11 (all images), 14 (left and right), 15 (left and right), 16 (left and right), 17 (left and right), 19 (left and right), 20 (left and right), 25 (left and right), 28–9 (all images), 35 (left and right), 37 (all images), 38 (left and right), 39 (left and right), 40 (left and right), 41 (left and right), 42 (left and right), 45 (left and right), 46 (left and right), 47 (all images), 48–9 (all images), 51 (left and right), 52–3 (all images), 54 (left and right), 60–1 (all images), 62 (left and right), 63 (all images), 66 (left and right); John Craven © Merehurst Limited: p 65 (left and right); M. Hanks: p 89 (right); Murdoch Books Picture Library: 21 (left), 23 (right), 31 (right);
Harry Smith Collection: pp 12, 13 (above and below), 72 (left and right), 74, 77, 81 (left and right)

PUBLISHERS' NOTE
The publishers would like to thank Burnham Nurseries, Devon for kindly allowing photography to take place.

FRONT COVER: Dendrobium *'Trakool Red'*
TITLE PAGE: *Encyclia radiata*

Murdoch Books UK Ltd
Ferry House, 51–57 Lacy Road,
Putney, London, SW15 1PR
Tel: +44 (0)20 8355 1480
Fax: +44 (0)20 8355 1499
Murdoch Books UK Ltd
is a subsidiary of
Murdoch Magazines Pty Ltd

UK Distribution
Macmillan Distribution Ltd
Houndsmills, Brunell Road,
Basingstoke, Hampshire,
RG1 6XS
Tel: +44 (0)1256 302 707
Fax: +44 (0)1256 351 437
http://www.macmillan-mdl.co.uk

Murdoch Books®
GPO Box 1203, Sydney,
NSW 1045, Australia
Tel: +61 (0)2 8220 2000
Fax: +61 (0)2 8220 2020
Murdoch Books®
is a trademark of
Murdoch Magazines Pty Ltd